SpringerBriefs in Energy

For further volumes:
http://www.springer.com/series/8903

Zafrullah Jagoo

Tracking Solar Concentrators

A Low Budget Solution

 Springer

Zafrullah Jagoo
Department of Physics
University of Mauritius
Réduit
Mauritius

ISSN 2191-5520   ISSN 2191-5539   (electronic)
ISBN 978-94-007-6103-2   ISBN 978-94-007-6104-9   (eBook)
DOI 10.1007/978-94-007-6104-9
Springer Dordrecht Heidelberg New York London

Library of Congress Control Number: 2012955751

Printed on acid-free paper

Springer is part of Springer Science+Business Media (www.springer.com)

"Every morning in Africa, a Gazelle wakes up. It knows it must run faster than the fastest Lion or it will be killed.

Every morning a Lion wakes up. It knows it must outrun the slowest Gazelle or it will starve to death.

It doesn't matter whether you are a Lion or a Gazelle: when the sun comes up, you'd better be running."

—Anonymous

To my beloved parents,
Osman and Zaheda

Foreword

The past decade has been one of the sustained growth as far as the market penetration of solar thermal and solar photovoltaic technologies in Europe, Japan, the USA, China, and India are concerned. Though developing countries of the African continent possess abundant solar resources, they have been largely left out of this revolution, mainly due to factors like poor institutional framework, lack of qualified staff, and prohibitive cost of the technology. Paradoxically, about 65 % of sub-Saharan Africa's population does not have access to electricity and the power requirements of its growing population are increasing rapidly.

There is, therefore, an urgent need for sub-Saharan African countries to tap indigenous renewable resources through technology transfer as well as through the provision of education, training, and research facilities for their sustainable development. These will have to be financed by a combination of carbon taxes, foreign aid, and government schemes for the promotion of renewable energies.

Zafrullah Jagoo's book presents a detailed account of the design of a Tracking Solar Concentrator. It describes the construction of a parabolic solar concentrator and exploits the power of a microcontroller device for automatically steering it toward the Sun at all times. Throughout this work, there is a deliberate attempt to use low cost and easily available components, materials, and tools. Moreover, the design is modular, so that some elements (electronics, mechanical structure, collector) can readily be changed to suit particular applications.

November 2012   Shailendra Oree

Preface

Various renewable energies have existed well before the apparition of human beings, yet we have failed to harness their true potential. The bottleneck with renewable energy systems is that they are extremely expensive to initiate and their production capacity is much less when compared to fossil-fuels. The primordial aim in publishing this book is tackle the first problem of high commencing cost by carving the path to a cheap but powerful, precise, and robust tracking solar concentrator. Since solar energy is the most abundant form of renewable energy, often present for days without faltering, it was the default starting choice.

The exorbitant cost of solar trackers is tackled from its roots. To ensure that the tracking mechanism was affordable to the general public, a very limited budget was allocated to building the steerable solar tracker. To maintain low price, an investigation was first carried out without leaving any stones unturned and then a blueprint of the solar tracker was implemented. To shift the design from paper to the real world, the skeleton of the system was first built followed by the gearing mechanics which was built from scratch using components instead of embedded commercial systems.

After the generic solar tracker had been conceptualized, a solar concentrator which helps focussing the light of the sun to a minuscule point was analyzed. The design phase and subsequently the implementation part took over. Emphasizing on inexpensive raw materials, a concentrator was made from fibre glass which has both the advantages of being malleable and low-priced. To really focus sun energy, the concentrator was lined up with a mirror film.

The only way to ascertain that our proposed solution is a viable option was to thoroughly test the machine in the field. It was shown that the tight-budget robot was successful in resisting the harsh conditions prevailing outside and proved to be a decent contestant in the commercial world in terms of cost, power, and reliability if the system is scaled up.

This work was written with three kinds of reader in mind:

- *general readers* who wants a treatment of the field that has both breadth and technical depth;
- *teachers and students* who want an authoritative text that covers all areas of solar concentrator and solar tracking;
- *researchers and engineers* who are interested in introductory treatments of advanced topics and also are interested in gaining expertise in this area and want a practical guide with some theoretical basis.

  Being an all-audience manuscript, no prior know-how is expected on the part of readers. All relevant equations have been derived from first principle and all notions have been elaborated using simple comprehensive analogies.

This research bears the imprints of many people who have contributed in one way or the other in its completion.

First of all I wish to express my thanks and gratitude to Dr. S. Oree for his unrelenting attention, excellent guidance, and comments while reviewing my book. Project manager, Dr. Oree initiated the theme of solar concentrators and encouraged me to start an ongoing research in the field of renewable energy. I would also like to thank Dr. G. Beeharry who helped me with microcontroller programming and to thank Dr. R. Somanah who agreed to support me while I was embarked on the research journey.

Next, to the wonderful folks at Springer Earth Sciences and Geography: Petra van Steenbergen who believed in my book and gave me fruitful advices; Hermine Vloemans who ensured that the process is as smooth as possible and who never got tired of replying my emails; and least but not last, Christian Witchel who agreed to endorse my book.

A special thank goes to all those people not mentioned but have in one way or another supported me to achieve this work of art. I would like to thank my mother and my father for their moral, spiritual support, and ongoing kindness throughout this project.

Last but not least, I am grateful to God who has sorted me out of every difficult situation.

I hope that you enjoy reading this book as much as I enjoyed researching and writing it. If you have suggestions or comments about this book or would even like to glimpse on my algorithms/scripts, please do not hesitate to shoot me an email (*zaf@physicist.net*).

Mauritius, November 2012   Zafrullah Jagoo

Contents

Acronyms

BCD	Binary coded decimal
CNC	Computer numeric control
DST	Daylight saving time
EEPROM	Electrically erasable programmable read-only memory
IC	Integrated circuit
MOSFET	Metal-oxide semiconductor field effect transistor
NVRAM	Non-volatile random-access memory
PCB	Project circuit board
PPM	Parts per million
PV	Photovoltaic
RAM	Random-access memory
RTC	Real-time clock

Symbols

Symbol	Description	Numerical value
α	Newtonian constant	J/°Cs
δ	Declination	
Δd	Diameter of Sun's image	m
ρ_T	Energy per unit time per unit emitter surface area per unit wavelength	J/m^{3}s
σ	Stefan–Boltzmann constant	5.67×10^{-8} W/m^{2}K^4
θ	Temperature	°C
θ_o	Room temperature	°C
$\dot{\theta}$	Rate of change of temperature	°C/s
alt	Altitude	
az	Azimuth	
A	Area	m^2
c	Speed of light in vacuum	3.0×10^8 m/s
d_{E-S}	Earth–Sun's distance	149,597,870,000 m
d	Diameter of parabolic dish	m
D	Diameter of load	m
e	Eccentricity	
E	Eccentric anomaly	
E_o	Surface irradiance	W/m^2
f	Focal length	m
g	Mean anomaly	
c_p	Specific heat capacity at constant pressure	4180 J/kg°C
h	Plank's constant	6.62607×10^{-34} Js
H	Hour angle	
$I_\odot$	Irradiance	W/m^2
JD	Julian Date Number	
k	Boltzmann's constant	1.38×10^{-23} J/K
lat	Latitude of Mauritius	$-20.28°$ S
$long$	Longitude of Mauritius	$+57.55°$ E
L	True Longitude	

(continued)

(continued)

Symbol	Description	Numerical value
$L_\odot$	Luminosity	W
$L0$	Altitude of Sunrise	
LST	Local Sidereal Time	
$LST0$	Local Sidereal Time at 0 hr Local Time	
m	Mass	kg
OE	Obliquity of ecliptic	
P	Power	W
$\dot{Q}$	Rate of change of energy	J/s
r	Earth–Sun's distance	1.50×10^{11} m
R	Radius	m
R'	Fraction of energy lost to reflection	
$R_\odot$	Radius of the Sun	6.96×10^8 m
RA	Right ascension	
t	Time	s
$T_\odot$	Temperature of the Sun	5777 K
UT	Universal Time	
v	True anomaly	
w	Argument of perihelion	

Chapter 1
Introduction

Abstract Harnessing a multitude of complementary green energy sources is the only plausible way to satisfy the energy demands of a greedy global economy. The potential of solar energy (being the most abundant) in fulfilling part of the energy requirements of mankind is immense and constitutes the focal point of this book. A self-powered solar tracker that points directly towards the sun thanks to an integrated control mechanism with two degrees of rotational freedom was studied and developed. The electro-mechanical control system is based on a precisely-timed microcontroller circuit that first computes the altitude and azimuth of the sun in real-time and then drives a pair of stepper motors that steers the solar tracker towards it. An indigenously built fibre-glass parabolic dish, whose surface was lined with a reflective vinyl mirror film served to concentrate sun rays incident on its surface.

We, human beings can't help but ask ourselves how and why things occur. There are so many questions, some answered and some still unanswered. If only I could turn my head towards the sky and seek the answers from that one silent observer who has witnessed it all, the sun, our Sun.

The use of concentrated energy from the sun dates back to around 700 B.C. when the Chinese used, for the first time ever, "burning mirrors" to ignite firewood. Aeons later, in his concern about mankind's destruction of nature and the environment, Leonardo Da Vinci laid the foundations for solar concentrators—an invention having the beauty of simplicity. He scribed in his notebooks the explicit designs which have inspired many since then to implement a near-perfect solar concentrator (Romm 2010).

1.1 Why so much Emphasis is Being Laid on Renewable Energies?

In July 2008, a barrel of crude oil at just over $145 (U.S. Energy Information Administration 2009) sufficed to highlight the inadequacy between oil reserves and the greed

Z. Jagoo, *Tracking Solar Concentrators*, SpringerBriefs in Energy, DOI: 10.1007/978-94-007-6104-9_1, © The Author(s) 2013

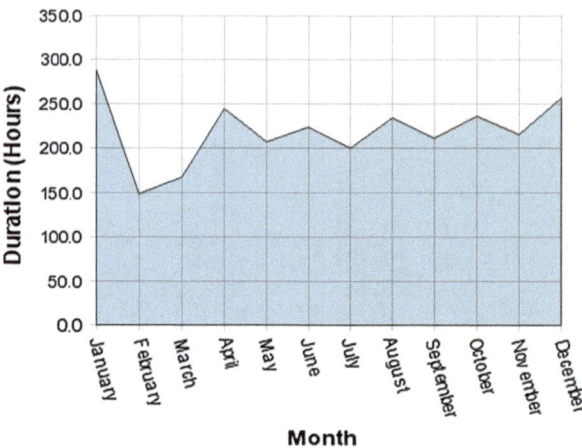

Fig. 1.1 Duration of *bright* sunshine for the year 2005 (*Source* Mauritius meteorological services)

with which the world economy is devouring this scarce resource. To make matters worse, burning these fuels increases the already present greenhouse gases leading to global warming. The fact that the amount of solar energy reaching the earth in more or less an hour is enough to power the globe for the whole of a year is thought provoking and fully justifies the attempts to recover solar energy. Harnessing this free energy at high efficiencies is considered, even today, as a challenge to engineers.

Solar concentrator technology with its inherent ability to produce the high temperatures required for efficient heat engines is, in my opinion, predestined to a bright future given the amount of research presently devoted to its development. However future successes rely on continuous and sustained efforts over a long term.

1.2 What is the Scope for Solar Energy Usage in Mauritius?

The total duration of sunshine in Mauritius on a monthly basis for the year 2005, from data collected at Mauritius Meteorological Services, is depicted in Fig. 1.1. On average, the island received around 8 h of bright sunshine daily, save February and March because of three tropical depressions and a severe tropical cyclone roaming in the neighbourhood of Mauritius. Solar energy is so abundant that recovering a fraction of this green energy will surely lead to an economically and ecologically sound future. The amount of primary energy consumed in Mauritius yearly is about 1.6 million tons of oil equivalent (Mtoe) or about 18.6 TWh. This is roughly the amount of solar energy incident on 14 km^2 of flat land (Palanichamy et al. 2004). The challenge is to convert, store and use this energy bestowed upon us by Mother Nature as effectively and efficiently as we can.

1.3 Purpose of Book

This project aims at constructing a full-fledged solar concentrator capable of generating a high solar flux density at its focus. The working model should require minimum human intervention and support, be low cost and made with widely available hardware. The set-up deployed is a lab-scale model which in due course can be made even more robust to withstand the fierce weather conditions prevailing in the outside.

It is however beyond the scope of this work to speculate on issues of large-scale implementation and its economics.

1.4 Book Structure

An outline of the remaining chapters of this book is as follows:

Chapter 2 discusses the different aspects of the sun's motion and addresses energy related issues. Moreover, a clear and concise primer on the position of the sun with respect to a particular position on the earth's surface is given.

Chapter 3 justifies the need for tracking in a concentrating solar power system. It also gives a review of the existing solar trackers that have been marked as reliable. It also details each stage of the design of our own lab-scale model of the solar tracker until a fully functional set-up has been obtained.

Chapter 4 overviews the numerous types of solar concentrators available throughout the globe. It also examines the elements of solar concentrators and the purpose for which they are built. The different step-by-step procedures towards the realization of our concentrating surface is enumerated.

Chapter 5 displays the results from several experiments carried out in the open spanning over several weeks. It also evaluates the performance of the actual system.

Chapter 6 concludes the book by summarizing the important results and also suggesting future recommendations to improving the prototype, thus scaling it to fit the outside scenery.

References

Palanichamy C, Babu NS, Nadarajan C (2004) Renewable energy investment opportunities in Mauritius—an investor's perspective. Renew Energy 29(5):703–716

Romm J (2010) Straight up: America's fiercest climate blogger takes on the status quo media, politicians, and clean energy solutions. Island Press, Washington, DC

US Energy Information Administration (2009) Annual energy review. Energy Information Administration, Washington, DC

Chapter 2
The Physics of the Sun

Abstract The chapter starts by describing briefly the basic features of sun and then proceeds to deriving the relevant equations that allow the calculation of several parameters pertaining to the sun namely the eccentric anomaly, hour angle, and the position of the sun (azimuth and altitude) amongst others. The relevant formulae are valid at any spacial and temporal location on the earth. The rule which allows us to compute the sun's position at any time is cross-checked with the US National Renewable Energy Laboratory's Solar Position Algorithm and is shown to be in good agreement ($<1\%$) with the exact sun's position at our present location ($20°17'$ S and $57°33'$ E).

The sun is a huge ball of hot gas subject to the action of gravitational forces that tend to make it shrink in size. This force is balanced by the pressure exerted by the gas, so that an equilibrium size prevails. The core of the sun which extends from the centre to about 20 % of the solar radius is at an extremely high temperature of around 15.7×10^6 K and pressure of 340 billion times earth's air pressure at sea level. Under these extremes, a nuclear fusion reaction takes place that merge four hydrogen nuclei or protons into an α-particle (helium nucleus), resulting in the production of energy from the net change in mass due to the fact that the alpha particle is about 0.7 % less massive than the four protons. This energy is carried to the surface of the sun in about a million years, through a process known as convection, where it is released as light and heat (Hufbauer 1991). Figure 2.1 shows the internal structure of the sun: the radiative surface of the sun, or photosphere, is the surface that emits solar radiation to space and has an average temperature of about 5,777 K. Localized cool areas called sunspots occur in the photosphere. The chromosphere (around 10,000 K) is the region where solar flares composed of gas, electrons, and radiation erupts. The corona forms the outer atmosphere of the sun from which solar wind flows (Mullan 2009).

Z. Jagoo, *Tracking Solar Concentrators*, SpringerBriefs in Energy,
DOI: 10.1007/978-94-007-6104-9_2, © The Author(s) 2013

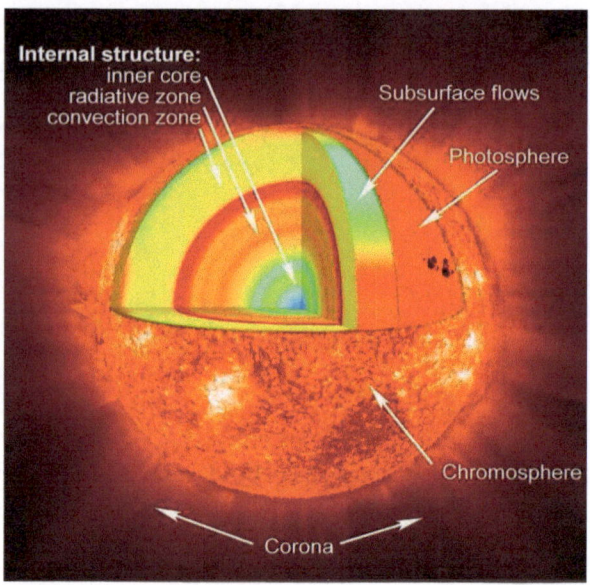

2.1 Irradiance and the Electromagnetic Spectrum

In first approximation, the sun can be considered to be a black-body emitter.
Figure 2.2 shows a comparison between the solar spectral irradiance incident at the
top of the Earth's atmosphere and the spectral irradiance of a black-body source at a
temperature of 5,777 K.

By and large, the spectra are similar. One noteworthy point is that the solar spectrum is interspersed with atomic absorption lines from the tenuous layers above the
photosphere.

Valuable estimates of the solar irradiance are obtained from Plank's Law:

$$\rho_T(\lambda)d\lambda = \frac{8\pi hc}{\lambda^5} \frac{d\lambda}{e^{\frac{hc}{\lambda k\theta}} - 1} \tag{2.1}$$

where

ρ_T = energy per unit time per unit emitter surface area per unit solid angle per
unit wavelength (J/m^3 − s),

h = Plank's constant (Js),

c = velocity of light in vacuum (m/s),

k = Boltzmann's constant (J/K),

θ = temperature of the object radiating energy (K), and

λ = wavelength of the radiation (m).

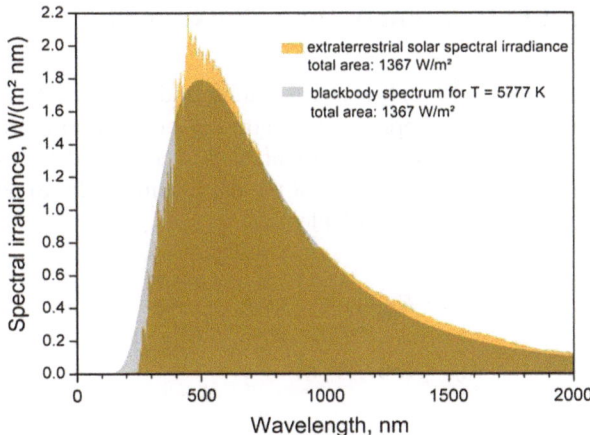

Fig. 2.2 Spectral irradiance of the sun (*Source* http://commons.wikimedia.org/wiki/File:
EffectiveTemperature_300dpi_e.png)

The wavelength of radiation λ_{max} at which peak emission occurs from a black-body is dependent upon the temperature of the object and can be calculated using the Wien's-Displacement Law derived from Plank's energy spectrum,

$$\lambda_{max}\theta = 0.2014\frac{hc}{k} = 2.898 \times 10^{-3} \text{ mK.} \tag{2.2}$$

Taking the average surface temperature of the sun to be 5,777 K, we compute $\lambda_{max} = 501$ nm, which is in the yellow-green region of the visible spectrum.

The irradiance E (W/m^2) or total amount of energy per second emitted per unit area of the black-body at temperature θ can be evaluated from Stefan-Boltzmann's Law:

$$E_o = \sigma\theta^4 \tag{2.3}$$

where
   E_o = surface irradiance of the object (W/m^2),
   $\sigma = 5.67 \times 10^{-8}$ W/m^2 K^4 is the Stefan-Boltzmann constant.

For the sun, we compute $E_\odot = 6.33 \times 10^7$ W/m^2. Over the whole surface area of the sun, $A_\odot$, the amount of radiated power per second called the solar luminosity $L_\odot$ is:

$$L_\odot = E_\odot \times A_\odot = E_\odot \times 4\pi R_\odot^2 = 3.85 \times 10^{26} \text{ W} \tag{2.4}$$

where
   $R_\odot$ = radius of the sun (m).

2.2 Theoretical Estimation of the Solar Constant

Solar radiation travelling through a near vacuum and reaching the top of earth's atmosphere is not subject to appreciable scattering or absorption on its path. We define the solar constant $I_\odot$, as the amount of power carried by incoming solar radiation (measured at the outer surface of the earth's atmosphere) per unit surface area perpendicular to the sun rays (Stickler 2003).

$$I_\odot = E_\odot \frac{A_\odot}{A_{sphere}} = E_\odot \frac{4\pi R_\odot^2}{4\pi d_{E-S}^2} \tag{2.5}$$

$$I_\odot = 1367 \text{ W/m}^2$$

where
d_{E-S} = earth-sun's distance—1 AU (m),[1]
$I_\odot$ = irradiance at a distance of 1 AU (W/m^2),
A_{sphere} = surface area of a sphere of radius 1 AU.

The radiant power available at the surface of the Earth's crust is lower than the solar constant due to a variety of factors, the main ones being (Batey 1998):

- reflection from clouds—cloud cover is one of the main factors blocking the rays of the sun and the amount of radiation reaching the earth on a cloudy day is diminished drastically as compared to a sunny day.
- atmospheric absorption—atmospheric aerosols (ozone, dust layer, air molecules, water vapour etc.) absorb selectively parts of the solar spectrum. A beneficial aspect of this effect is that it prevents destructive ultra-violet rays from damaging our health.
- cosine effect—at high latitudes, sunlight is incident on a level ground at large angles of incidence after travelling through a thick layer of atmosphere. where it undergoes considerable scattering and absorption. This accounts for the reduction in the available solar power, particularly in winter even if the receiving surface is perpendicular to sun rays.

On a very clear day, atmospheric absorption and scattering of incident solar energy cause a reduction of the solar input by about 22 % to a maximum of 60 % (US Department of Energy 1978).

---

[1] An Astronomical Unit (AU) is approximately the mean distance between the Earth and the Sun and is equal to 149,597,870,000 ± 6 m.

2.3 Astronomy

2.3.1 The Horizontal (alt–az) System

The horizontal coordinate system as a celestial coordinate system is most immediately related to the observer's impression of being on a flat plane (local horizon) and at the centre of a vast hemisphere across which heavenly bodies move (cf. Fig. 2.3). An observer in the southern hemisphere can define the point directly opposite to the direction in which a plumb-line will hang as the zenith (Bhatnagar and Livingston 2005). There are two coordinates that specify the position of an object in this system:

1. Altitude angle (Alt) or elevation is the angle between an imaginary line from the observer to the sun and the local horizontal plane (shown in 'green' in Fig. 2.3).

   - when the sun is "above the horizon": $0° \leq Alt \leq 90°$
   - when the sun is "below the horizon": $-90° \leq Alt \leq 0°$

2. Azimuth angle (Az) is the angle measured clockwise between the northern direction and the projection on horizontal ground of the line of sight to the sun (shown in 'red' in Fig. 2.3). The azimuth ranges from 0 to 360°.

The advantage of the alt–az system is its simplicity to take measurements as only one reference point (North) is needed. However, the main disadvantage is that this system is purely local and observers at different locations on the earth will measure different altitudes and azimuths for the same celestial object even though the measurements are made at the same time.

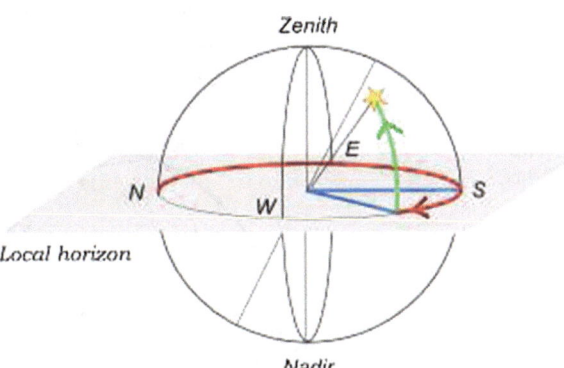

Fig. 2.3 Horizontal coordinate system (*Source* http://en.wikipedia.org/wiki/File:Horizontal_coordinate_system_2.png)

2.3.2 The Equatorial System

The equatorial coordinate system is used to illustrate the motion of heavenly stars on the celestial sphere—an imaginary sphere of radius equal to the distance of stars so that they appear to be lying on its surface. The projection of the earth's equator onto the celestial sphere is called the celestial equator. Similarly, projecting the geographic poles onto the celestial sphere defines the North (NCP in Fig. 2.4) and South (SCP in Fig. 2.4) celestial poles which maintain a relatively fixed direction (in the lifetime of a person) with respect to the distant stars.

Over a much longer term, the earth's rotation axis precesses about the ecliptic North Pole with a period of 25800 years. As a consequence, the North celestial Pole presently pointing towards Polaris will point towards Vega after half the precession period and back towards Polaris after one precession period. Owing to the daily rotation of the earth, the Greenwich Meridian sweeps across the celestial sphere and thus cannot be used as reference for locating stars. Instead, the meridian (longitude) of the vernal equinox is used as the zero celestial meridian. Equinoxes occur twice yearly and correspond to positions of the sun lying in the plane of the celestial equator. The first of the equinoxes, the vernal equinox occurs around March 21 with the sun oriented towards Pisces constellation. Due to the slow precession of the equinoxes, there is a small westward deviation in the direction of the vernal equinox by 50 arc seconds yearly.

The equatorial coordinates of a star are (Bhatnagar and Livingston 2005):

1. Declination (δ) is the angular distance of the sun north or south of the earth's equator. It is analogous to the latitude on planet earth, extrapolated to the celestial sphere. The earth's equator is tilted $23°27'$ with respect to the plane of the earth's orbit around the sun, so at various times during the year, as the earth orbits the sun, declination varies from $+23°27'$ (north) to $-23°27'$ (south). This change in the

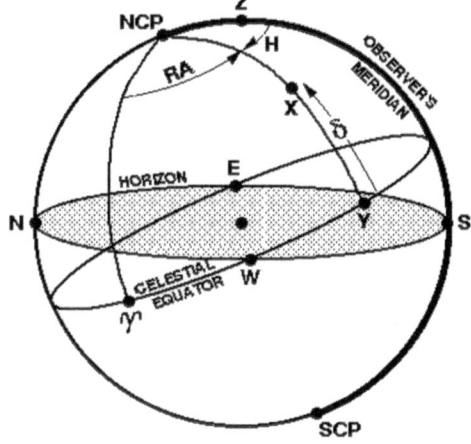

Fig. 2.4 Equatorial coordinate system (*Source* http://www.vikdhillon.staff.shef.ac.uk/teaching/phy105/celsphere/equatorial.gif)

value of declination is responsible for seasonal changes. For more information, please refer to pp.73 in *Astronomy: Principles and Practice* by Roy and Clarke (2003).

2. Right Ascension (*RA*) is measured in hours, minutes and seconds east from the meridian of the vernal equinox (zero meridian) to the star's meridian or hour circle. The hour circle is the great circle that passes through the poles and the stars, that is, right ascension is the time interval between the most recent overhead passage of the meridian of the vernal equinox and the overhead passage of the hour circle. As declination is analogous to latitude on the earth, so is right ascension to longitudes. Alternatively, hour angle can be used in place of *RA*. The hour angle (*H*) indicates the time elapsed since the star transited across the local meridian. Although it is calculated from measurements of time, it may be expressed in angular units (Roy and Clarke 2003).

Unlike the horizontal system, equatorial coordinates do not depend on the observer's location. As a matter of fact, only one pair of coordinates is required for an object at all times.

2.4 The Sun's Position

The change in coordinates of the sun is brought about by a plethora of factors. The first and most obvious motion of the sun is the daily rotation about its north–south axis. The second is a seasonal north–south motion of $\pm 23°27'$ away from the equator. The third motion is a subtle change in the sun's noontime position, brought on mostly by the earth's axial tilt, but with a small additional component produced by the earth's non-circular (elliptical) orbit around the sun.

Since we are to harvest the energy from the sun, it is imperative to know the sun's position at any time. Our formulae should comprise of all the three types of behaviour of the sun. To do so, we will first compute the Julian day number from J2000.0 epoch,

$$JD = 367 \times \text{INT} \left[\frac{7 \times \text{INT} \left[\frac{(M+9)}{12} \right]}{4} \right] + \text{INT} \left[275 \times \frac{M}{9} \right] + D - 730530 + \frac{UT}{24}$$

(2.6)

where,

D is the calendar date,

M is the month,

Y is the year and

UT is the Universal Time in hours only, i.e. the local time less 4 h if daylight saving time (DST) is not applicable and less 5 h if DST is on. INT[ ] is a function that discards the fractional part and returns the integer part of another function. The following parameters may be calculated directly from the Julian date.

$$\text{Obliquity of eliptic, } OE = 23.4393 - 3.563 \times 10^{-7} \times JD \qquad (2.7)$$

$$\text{Argument of perihelion, } w = 282.9404 + 4.70935 \times 10^{-5} \times JD \qquad (2.8)$$

$$\text{Mean anomaly, } g = 356.0470 + 0.9856002585 \times JD \qquad (2.9)$$

$$\text{Eccentricity, } e = 0.016709 - 1.151 \times 10^{-9} \times JD \qquad (2.10)$$

We can henceforth formulate the eccentric anomaly in degrees as:

$$E = g + e \times \left(\frac{180}{\pi}\right) \times \sin{(g)} \times (1 + e \times \cos{(g)}). \qquad (2.11)$$

Next, we will give the expression for the true anomaly, v and the earth-sun's distance, d.

$$d \times \sin{(v)} = \sqrt{1 - e^2} \times \sin{(E)} \text{ and} \qquad (2.12)$$

$$d \times \cos{(v)} = \cos{(E)} - e. \qquad (2.13)$$

The sun's true longitude, L can now be computed:

$$L = v + w. \qquad (2.14)$$

Since the sun is always at the ecliptic (or extremely close to it), we can use simplified formulae to convert L (the sun's ecliptic longitude) to equatorial coordinate systems:

$$\sin{(\delta)} = \sin{(OE)} \times \sin{(L)} \qquad (2.15)$$

$$\tan{(RA)} = \frac{\sin{(L)} \times \cos{(OE)}}{\cos{(L)}}. \qquad (2.16)$$

In earth's frame, the alt–az system is easier to implement. We make use of the following relations to compute alt and az from RA and δ. Since on earth, it is easier to work with azimuth and elevation of an object, we need to compute the azimuth and elevation but first we must compute the Local Sidereal Time (LST) of the place and time in question in degrees,

$$LST = 98.9818 + 0.985647352 \times JD + UT \times 15 + lon. \qquad (2.17)$$

where lon is the longitude of the observer.
We compute the Hour Angle (H),

$$H = LST - RA. \qquad (2.18)$$

Now with all the data available, we can compute Alt and Az by applying the cosine formula for spherical geometry:

$$Alt = \sin^{-1}\left[\sin(lat) \times \sin(\delta) + \cos(lat) \times \cos(H)\right] \qquad (2.19)$$

$$Az = \tan^{-1}\left[\frac{\sin(H)}{\cos(H) \times \sin(lat) \times \tan(\delta) \times \cos(lat)}\right] + 180 \qquad (2.20)$$

where lat is the latitude at the observer's location. To compute when a star rises, we must first calculate when it passes the meridian and the hour angle of rise. Apart from the problem of calculating the sun's position in space relative to earth, one must also calculate the relative motion of the sun at each point on the earth's surface.

To find the meridian time, we compute the Local Sidereal Time at 0 h local time as outlined above and we name that quantity LST0. The Meridian Time, MT, will now be:

$$MT = RA - LST0 \qquad 0° \le MT \le 360°. \qquad (2.21)$$

Now, we compute H for rise, and we name that quantity $H0$:

$$\cos(H0) = \frac{\sin(L0)\sin(lat)\sin(\delta)}{\cos(lat)\cos(\delta)} \qquad (2.22)$$

where $L0$ is the selected altitude selected to represent sunrise. The sun would normally appear to be exactly on the horizon when its altitude is zero, except that the atmosphere refracts sunlight when it's low in the sky, and the observer's elevation relative to surrounding terrain also impacts the apparent time of sunrise and sunset. The difference between the time of apparent sunrise or sunset and the time when the sun's altitude is zero is usually on the order of several minutes, so it's necessary to correct for these factors in order to obtain an accurate result. For a purely mathematical horizon, $L0 = 0$ and for a physical solution, accounting for refraction on the atmosphere, $L0 = -35/60°$. The effects of the atmosphere vary with atmospheric pressure, humidity, temperature etc. Errors in sunrise and sunset times can be expected to increase the further away from the equator, because the sun rises and sets at a very shallow angle. And if we want to compute the rise times for the sun's upper limb to appear grazing the horizon, we set $L0 = -50/60°$.

The rise time of the sun is given by:

$$Sunrise = MT - H0. \qquad (2.23)$$

The answer from the above equation will be in degrees and is should be converted to hours.

However, in everyday experience, the sunset time prediction is not 'usually' borne out as the atmosphere is disturbed, because one is almost never standing on a flat plain with barriers on the horizon (Duffet-Smith 1988; Meeus 1999).

Table 2.1 Table showing the exact values and the theoretical estimates of the solar position

Time	Theoretical azimuth	Theoretical altitude	Exact azimuth	Exact altitude
08 00	105.4°	32.13°	105.4°	32.01°
09 00	103.0°	45.76°	102.9°	45.67°
10 00	101.7°	59.51°	101.6°	59.41°
11 00	103.6°	73.26°	103.4°	73.16°
12 00	140.2°	86.04°	139.7°	86.01°
13 00	252.9°	78.24°	253.0°	78.36°
14 00	258.1°	64.80°	258.2°	64.68°
15 00	257.6°	50.80°	257.7°	50.91°
16 00	255.5°	37.12°	255.6°	37.23°
17 00	252.6°	23.58°	252.7°	23.70°
08 00	105.4°	32.01°	105.3°	32.01°
09 00	102.9°	45.65°	102.9°	45.55°
10 00	101.6°	59.40°	101.5°	59.30°
11 00	103.4°	73.15°	103.1°	73.06°
12 00	138.6°	86.01°	136.9°	85.97°
13 00	253.0°	78.36°	253.1°	78.48°
14 00	258.2°	64.91°	258.4°	64.79°
15 00	257.7°	50.91°	257.8°	51.02°
16 00	255.6°	37.22°	255.7°	37.33°
17 00	252.7°	23.68°	252.7°	23.80°

2.4.1 Validity of Sun's Algorithms

The final equations for the location of the sun were checked on 27 and 28 December 2008 by contrasting the calculated values from the derived equations in Sect. 2.4 to the exact values obtained from National Renewable Energy Laboratory's Solar Position Algorithm available at http://www.nrel.gov/midc/solpos/spa.html in Table 2.1. The latitude and longitude of observation were locked at 20°17′ S and 57°33′ E respectively. The complete script with all the relevant comments are shown in Appendix A.

2.4.1.1 Errors in Experiment

From the measurements, we can infer that the calculated solar position reflected the actual solar position most of the time (maximum error of 0.5 %). The algorithm could be enhanced to cater for errors at instants when the sun is in its maximum phase but increasing accuracy is both computer expensive and time-consuming, so a compromise between speed, resource and complexity yields our solar formulae.

2.5 Chapter Summary

The chapter starts by describing briefly the basic features of sun and then proceeds to deriving the relevant equations that allow the calculation of several parameters pertaining to the sun namely the eccentric anomaly, hour angle, and the position of the sun (azimuth and altitude) amongst others. The relevant formulae are valid at any spacial and temporal location on the earth. The rule which allows us to compute the sun's position at any time is cross-checked with the US National Renewable Energy Laboratory's Solar Position Algorithm and is shown to be in good agreement ($<1\%$) with the exact sun's position at our present location ($20°17'$ S and $57°33'$ E).

References

Batey M (1998) Spectral characteristics of solar near-infrared absorption in cloudy atmospheres. J Geophys Res 103(D22):28–793

Bhatnagar A, Livingston W (2005) Fundamentals of solar astronomy. World Scientific, Singapore

Duffet-Smith P (1988) Practical astronomy with your calculator, 3rd edn. Cambridge University Press, Cambridge

Hufbauer K (1991) Exploring the sun: solar science since Galileo. Johns Hopkins University Press, Maryland

Meeus J (1999) Astronomical algorithms, 2nd edn. Willmann-Bell, Virginia

Mullan D (2009) Physics of the sun: a first course. Taylor & Francis, London

Roy A, Clarke D (2003) Astronomy: principles and practice, 4th edn. Taylor & Francis, London

Stickler G (2003) Solar radiation and the earth system. National Aeronautics and Space Administration. http://education.gsfc.nasa.gov/experimental/July61999siteupdate/inv99Project.Site/Pages/science-briefs/ed-stickler/ed-irradiance.html. Accessed 7 Feb 2009

US Department of Energy (1978) On the nature and distribution of solar radiation. Watt Engineering

Chapter 3
Solar Tracking

Abstract The various types of solar trackers are reviewed in this chapter along with their merits and disadvantages. It has been shown that in terms of the relative power output, a dual-axes tracker is the most efficient system available. The step-by-step construction of a novel dual-axes solar tracker, that points directly towards the sun thanks to an integrated sun tracking mechanism with two degrees of rotational freedom, is presented in this chapter. Each stage of the design, with explicit explanation of all the components, and realisation of the solar tracker is detailed. The electro-mechanical control system is based on a precisely-timed microcontroller circuit that first computes the altitude and azimuth of the sun in real-time and then drives a pair of stepper motors that steers the system towards it. The system will track the sun throughout the day and return to its default position for night-time stowing. The whole set-up can be constructed in about 6 months at a record price of \$118.81 for the electronics circuitry that any generic solar tracker can utilize and \$159.78 for a tailor-made prototype frame.

While the preceding chapter elaborated on the mechanics and dynamics of the sun at any point in space and time, this chapter uses that information to help in the selection process of the type of tracker that is most suitable for holding a physical entity. Once the type of system has been picked, the route towards achieving that goal will be illustrated within this chapter's sections. This part of the book has the actual design of the system at its focus.

3.1 Fixed System Versus Solar Tracker

A fixed system, as its name suggests, is an immobile system that is mounted so that it intersects sun rays for the bulk part of a given day. An example of a fixed system is the inclined plate that is usually located atop houses to capture energy from the sun. Although on average, this system performs reasonably well, it does not maintain optimal orientation with respect to the sun at all instants. It is desirable to have a system that is optimally oriented, always at normal incidence with regards to the sun, so as to squeeze out maximum solar energy for a given footprint.

On the other hand, a tracking system is a sophisticated device comprising of electronic control circuitry and mechanical elements that orient the collector towards the sun at all times from sunrise to sunset. There exist two main classes of tracking mechanisms, the single-axis tracker and the dual-axes tracker. As the name implies, the single-axis tracker has a single degree of rotational freedom about an axis which is approximately parallel to the earth's axis of rotation that allows it to follow the east-to-west motion of the sun during the day. Single-axis trackers are unable to follow the sun with absolute accuracy due to the seasonal variation of the tilt of the earth's equatorial plane with respect to the earth's orbital plane. Due to the presence of two degrees of rotational freedom, dual-axes trackers are capable of perfect alignment with the sun at all times, giving optimal performance year-round.

However, if we examine them in a mechanical way and taking into consideration Murphys law that "if anything can go wrong, it will", then it is possible that at some random time, the system stops abruptly. Unless the solar tracker stops in the centre position at the middle of the day, we are likely to be worse off than with a fixed solar arrangement. Trackers are less resistant to natural calamities, and will surely sail off if cyclonic gales get underneath the collector. For countries where episodes of high wind conditions occur during short bursts of time, a tracker is best-suited.

The enhancement in performance brought about by solar tracking compared to a fixed system is largely dependent on the latitude of the installation as well as the design. For example, parabolic solar concentrators cannot be imagined without a precision tracking mechanism in order to guarantee that sunlight is converged exactly on the load. A fixed flat-plate receiver in Mauritius (20°17' S) would be oriented facing the north at around 20° to the horizontal for good performance even during winter months when the available solar illumination is relatively low and its duration is shorter. Clearly in this case, a tracking system would bring significant performance improvement, particularly throughout summer. Cotfras et al. (2008) has shown that, out of 15 h of sunlight in Germany, a dual-axes tracker has a relative power output of 100 % for 9 consecutive hours, while a single-axis tracker maximizes the power output for 5 h and a fixed system shows that the power output is at its limit for only an hour. A bar-chart for the comparison between a dual-axes, a single-axis and a fixed system is shown in Fig. 3.1.

3.2 Operating Principles of Solar Trackers

3.2.1 Dynamic Trackers

Dynamic trackers are the simplest tracking system in the sense that they utilize matched solar photovoltaic cells: namely cadmium sulphide stereogram sensors which generate a differential signal whenever the orientation of the device is not

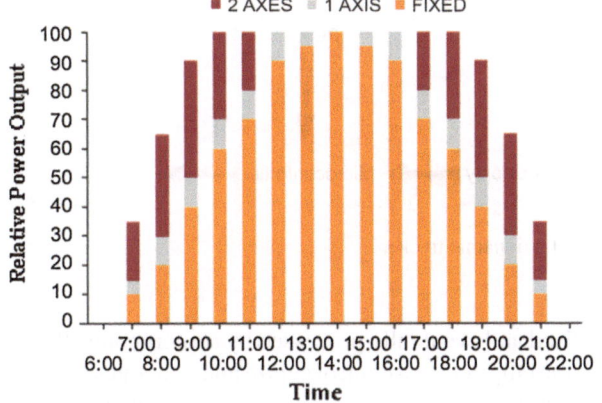

Fig. 3.1 Comparison between different tracking systems and a fixed system

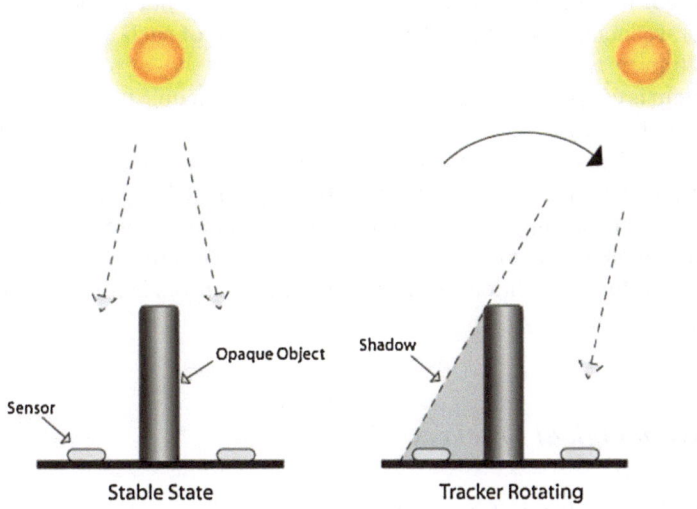

Fig. 3.2 CdS sensors for a dynamic tracker

optimal. Typically one pair of sensors is used for one-axis tracking to control a single motor and two pairs are required for the full-tracking mode (Mousazadeh et al. 2009).

One such system employs an embedded processor that is interfaced to the output of photo-voltaic (PV) cells or Cadmium sulfide (CdS) light sensitive resistors. As shown for a pair of cells in Fig. 3.2, an opaque pole is placed between a pair of PV cells, such that the shadow of the rod would fall on one of the two cells if the sun is not in the plane passing through the pole and perpendicular to the plane of the paper. This gives a differential output signal between the matched pair which is

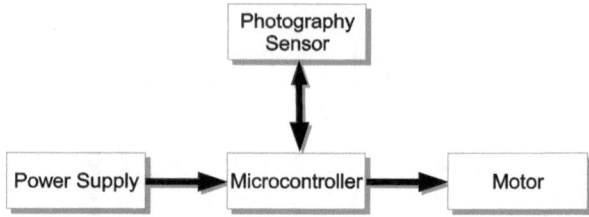

Fig. 3.3 Backbone of a dynamic tracker

amplified, digitalised and eventually processed by a microcontroller (DSP or FPGA). The controller in the chip acts as a control centre and generates the required signals to operate a stepper motor (via stepper motor drive circuitry) that changes the orientation of the axis that is perpendicular to the matched pair of PV cells (Huang et al. 2009). Two such matched pairs are implemented in dual-axes trackers, where the two axes are decoupled, i.e. the rotation angle of one motor does not influence that of the other motor, reducing control problems. The backbone of such a control device for a dynamic tracker is shown in Fig. 3.3.

The system is easy-to-build, control and maintain. Dynamic tracking devices have the difficulty of failing to discriminate between the obscured sun and a bright spot in a broken cloud. The feedback mechanism orientates the receiver towards the bright spot rather than the sun. Also, they may not always point at the centre of the sun as the readings of the sensors may not be different if the sensors are pointing at the edges of the sun rather than the centre. Furthermore, these devices are not dependable under foggy, misty or dusty conditions as the 'view' of the sensors is impaired, and hence tracking fails undoubtedly.

3.2.2 Chronological Trackers

A chronological tracker is frequently a single-axis device that employs a clock mechanism to maintain the receiver perfectly oriented in the direction of the star being followed. The axis of rotation of the tracker for a single-axis tracker is paralleled to the earth's axis, and the sense of rotation is east to west, i.e. opposite to that of the earth (Fig. 3.4). This is usually referred to as an equatorial mount. In such devices, the alignment of the tracker's axis is adjusted by hand. Receivers attached to the equatorial mount are oriented at an angle β with respect to the local horizontal plane. Due to seasonal variations, angle β varies progressively by $\pm 23.44°$ during a year, so the orientation of the receiver (β) needs frequent adjustments.

The chronological tracking system works only during the daytime over a known period of time when it rotates at a rate of 15° per hour. Microcontroller operation together with stepper motors are used for precise motion. At the end of the day,

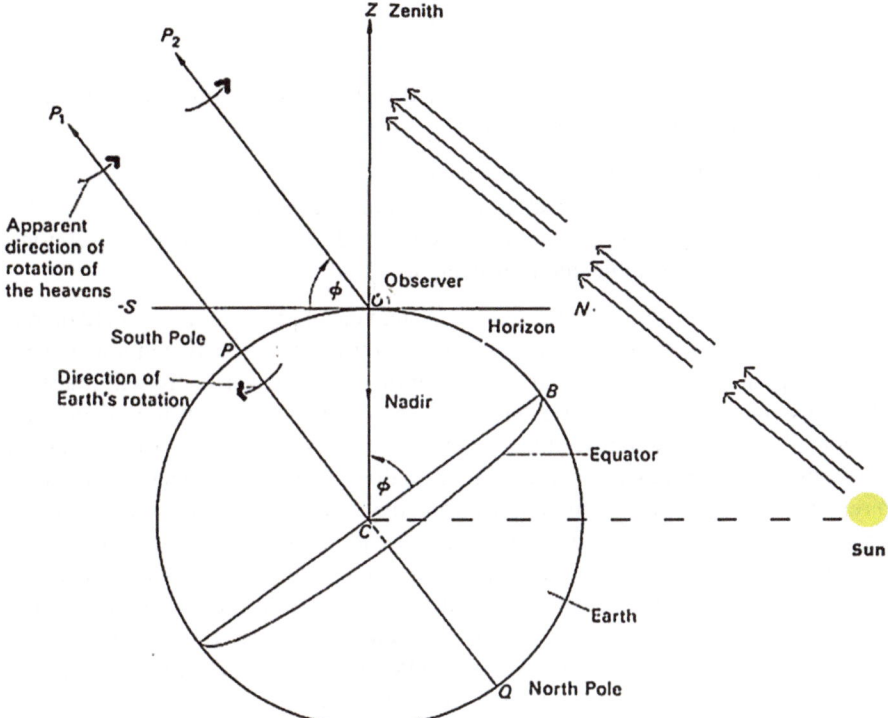

Fig. 3.4 Equatorial mount (in winter)

the tracker activates a limit switch which orders the microcontroller to bring the tracker back to default night-time position. At this stage, a timer is activated so that the mechanism is triggered back into action the following morning from a starting position that is set through the use of another limit switch. The range of angular motion can be adjusted by modifying the location of the limit switches. In the months of summer, the limiting range can be longer compared to that in winter (Barsoum 2011; Mousazadeh et al. 2009).

The system is cost-effective and relatively easy to implement and operate. Following the sun accurately in its daily as well as yearly seasonal motions is no easy task and a tracking of 15° per hour is not accurate. The use of the system over a long term imposes regular manual adjustments to the position of the tracker and manually changing the limit switches render tasks more complicated and an expert must be employed to stance the switches as a small variation in the position can lead to a big time difference, hence inaccurate tracking.

3.3 Overview of System Operation

The tracking system design is largely dictated by the expected operating conditions
in the field. During operation presumably,

- grid power would not be readily available.
- any point in space must be located with pin-point accuracy.
- the system would be working for extended periods without human intervention.
- system should be portable and light-weight.

These operating conditions imply that the system electronics should operate on battery power. To ascertain that the sun is followed during daytime, two motors have to
be placed perpendicular to each other. One motor oriented vertically will make the
whole system rotate in the $x-y$ plane while the other independent motor (pointing
horizontally) will make part of the system rotate in the $r-z$ plane. Additionally, to
reduce the extent of human intervention, a small solar PV unit is deemed necessary
to replenish the batteries continuously during daytime. If the power consumption of
the system is large, for maximum efficiency, the PV unit should be mounted on the
tracker itself, otherwise, the PV cells can be placed alongside the solar tracker.

The electronics control system should, by design, dissipate as little power as possible.
An embedded solution based on a microcontroller is therefore a very elegant result
though more complex to design compared to PC-based solutions. Added advantages
are the reduced cost and the small footprint and weight.

The microcontroller handles the burden of initialising the system as it is first powered, then computes and updates the parameter related angular orientation of the
system. A LED display interfaced to the microcontroller allows written output to
be displayed. The microcontroller also generates control signals for stepper motor
drive circuitry, which in turn powers the transistors attached to two stepper motors.
The stepper motors steer the tracker into optimal position with respect to the sun.
The microprocessor is helped in its function by a real-time clock, useful for precise
timing and storing the current time even when the system is off (like in a PC). A
schematic of the operation of the tracker is shown in Fig. 3.5.

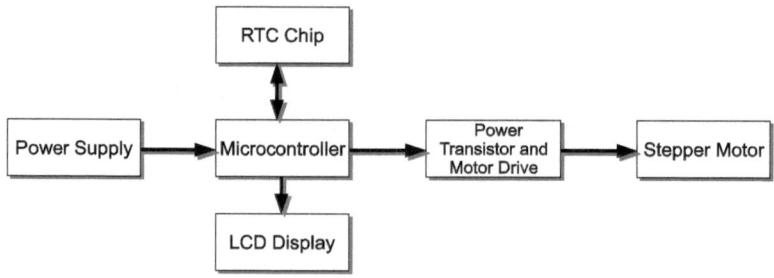

Fig. 3.5 Schema of the tracking system

3.4 Prototype Design

A prototype had to be implemented with the following characteristics;

Rigidity—Because the hardware is be exposed to rain, wind and temperature variations, it should be resistant to fierce weather conditions. Concrete, stainless steel or plated metals are possible choices. Unfortunately either wood or iron is not an alternative although both are cheap, as wood will rot once in contact with moisture just like iron will rust.

Ease of manufacture—The material required to build the system should be readily available. In addition, the tools required should be easy to find in a reasonably equipped workshop. Moreover, large-scale production which is just manufacturing independent components several times and putting them together in bulk should also be a viable alternative.

Cost—The bits and pieces used to build the prototype should not be expensive. The low-cost advantage will certainly make the solar tracker a tempting system.

3.5 Feasibility Study

A feasibility study, which investigates the viability of a solution with an emphasis on identifying potential problems, was carried out (British Computer Society 2002). As the goal of creating a dual-axes solar tracker carrying a concentrator has been set, we can take a look on the different factors that will affect the production of the tracking solar concentrator.

3.5.1 Technology/System Analysis

The technology for building solar concentrators is already well-grounded since solar concentrators are commercially available throughout the world but because of the monopoly and their scarcity, they are highly overpriced and certainly not suitable for small-scale production.

As far as the technical expertise is concerned, any skilled worker should be able to handle the completion of a solar concentrator efficiently once the blueprint is manufactured. Technical know-how is not limited to circuit prototyping, hardware assembly and soldering electronic parts.

3.5.2 Time-Factor Analysis

The time-factor feasibility estimates the time-frame that the system will take to be fully developed. If all the building blocks of the system—electric and electronic

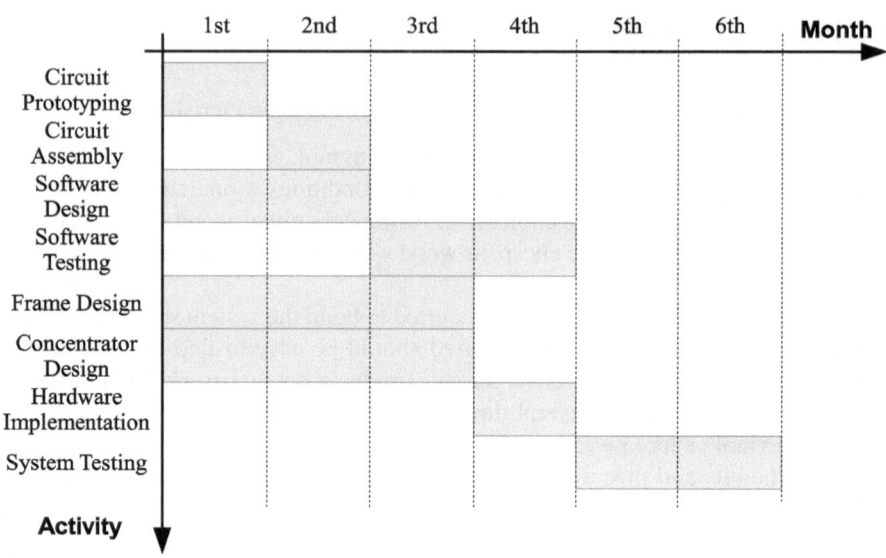

Fig. 3.6 Time plan of project

components, mechanical parts, frame construction parts and motors are available in time, then the system will be assembled in the required amount of time.

Figure 3.6 shows how the whole project is scheduled, which activity should be carried out first, followed by the next, and which activities should be done in parallel. From our study, it will take approximately to set-up a full-fledged solar concentrator from scratch if all the plans of the system are available.

3.5.3 Cost/Beneficial Analysis

One of the most important factors in any project is the ultimate cost of the system. A cost feasibility study gives an approximate idea of the cost that has to be incurred in implementing the new system and whether the proposed solution is cost effective, i.e. it can be completed with the allocated budget and if and only this is possible, then it is worth going ahead. Since the theme of this book is a low budget tracking solar concentrator, we imposed a low budget of $1,000 to the creation of both the solar tracker and the solar concentrator.

1. Cost incurred setting up the new system—This covers only the hardware costs because most of the software used are open source software (OSS) and are freely available over the internet.

   The cost given in Table 3.1 is an over-estimate of the real cost for a tracking solar concentrator prototype since tentative components have been used in this calculation.

Table 3.1 Cost feasibility study of tracking solar concentrator

Equipment	Credit ($)	Unit price[a] ($)	Debit ($)	Remaining balance($)
Budget	1,000			
Solar tracker				
Microcontrollers (X2)		5	10	990
Oscillators (X2)		1	2	988
Clock mechanism		4	4	984
Motor control IC (X4)		10	40	944
FETs (X8)		3	24	920
Microswitches (X2)		2	4	916
Stepper motor (X2)		112	224	692
Assorted resistors (X50)		0.5	25	667
Assorted capacitors (X50)		0.5	25	642
DIP sockets (X10)		0.5	5	637
PCB (X2)		10	20	617
Solar concentrator				
Stainless steel sheet (X2)		50	100	517
Aluminum tubes (X3)		20	60	457
Screws and bolts (X100)		0.5	50	407
Fibreglass (X3)		40	120	287
Fibreglass resin		35	35	252
Reflective film		40	40	212
Cement (X10)		4	40	172
Precision bearings (X20)		2	40	132
			Total cost	868

[a]Prices are obtained from Digi-Key Corporation (http://www.digikey.com/) and Home Depot (http://www.homedepot.com/)

2. Intangible Costs
   Intangible costs include all the costs that cannot be easily quantified in terms of rupees and cents. They may take the following forms:

   - the waiting time for the shipment to arrive,
   - the item posted may be swapped or defective,
   - the equipment may not be as durable as expected,
   - the problems that arise due to hardware failure.

The total cost of the system is barely less than the allocated budget which means that the project can proceed to the actual design and realisation.

28-Pin PDIP, SOIC

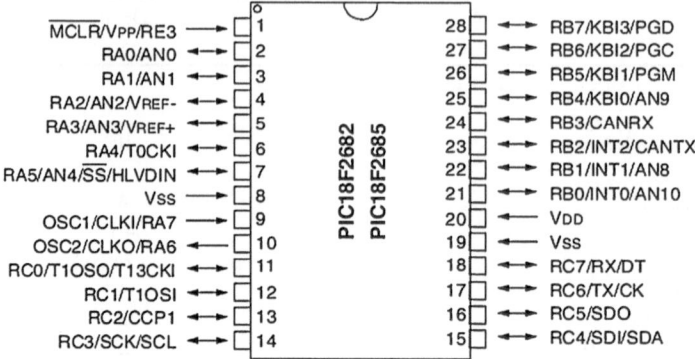

Fig. 3.7 Pin assignments of PIC18F2682®. Reproduced with permission from Microchip Technology Inc. (2008)

3.6 Solar Tracker Design and Realisation

Keeping the attributes listed in Sect. 3.4 in mind, we will describe the main components that are used to construct the tracker, and the way in which they interact with each other for its smooth operation in the remaining parts of this chapter.

3.6.1 Microcontroller

A microcontroller is a small computer furnished in a single integrated circuit and needing a minimum of support chips including a small amount of RAM, PROM, timers, and I/O ports to communicate with peripherals as well as complementary resources. Its principal nature is self-sufficiency, low power and low cost.

The PIC® microcontroller PIC18F2682® was selected to supervise and control the operations of the tracker. The PIC18® family offers a highly flexible solution for complex embedded applications. This family of devices offers the advantages of all high-speed microcontrollers namely, high computational performance at an economical price with the possibility of high-endurance, enhanced flash program memory (Microchip Technology Inc. 2008). The PIC18F2682® has 80 kilobytes of integrated flash memory for storage of written instructions, 3328 bytes of RAM to store variables while the processor is powered on, 1 kilobyte EEPROM to store stacks (address) and routines. It also has a 10-bit embedded analog to digital converter and communication components amongst which is the inter-IC serial communication. A summary of the electronic components of the PIC18F2682® is reported in Table 3.2 while a diagram of the actual chip is depicted in Fig. 3.7.

Table 3.2 PIC18F2682® specifications (Microchip Technology Inc. 2008)

Device	Program memory		Data memory		I/O	10 bit A/D (ch)	CCP1/ ECCP1 (PWM)	MSSP		EUSART	Comp.	Timers 8/16-bit
	Flash (bytes)	#single-word instructions	SRAM (bytes)	EEPROM (bytes)				SPI	Master I²C™			
PIC 18F2682	80K	40960	3328	1024	28	8	1/0	y	y	1	0	1/3

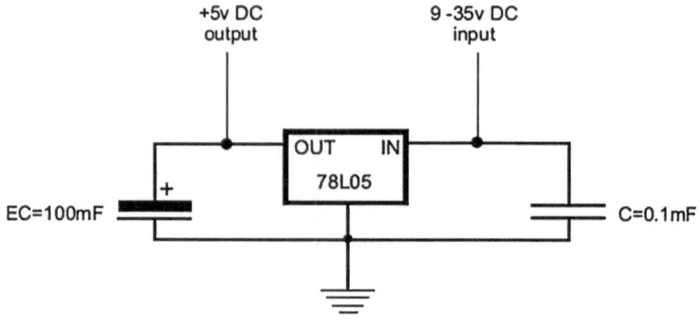

Fig. 3.8 Voltage stabiliser circuit for 7805

This processor has basically 3 ports (RA, RB and RC), each 8 output pins which can be controlled independently. Port C is mainly for communication purposes leaving only Ports A and B for input from or output to the real-world. Two pins from Port A are reserved for clocking the microcontroller as further explained in Sect. 3.6.1.2. Finally V_{ss} and V_{dd} are the ground reference and positive rail of the power supply respectively.

3.6.1.1 Powering Microcontrollers

The PIC18F2682® operates at 5.0 V DC which is provided by a 3-pin monolithic voltage regulator such as the 7805. The 7805 integrated circuit (IC) provides good regulation as well as automatic thermal shut-down and short circuit overload protection. Batteries usually have an emf of greater than 5.0 V, so the excess voltage causes heat dissipation at the metallic plate in the 7805. The circuit for the voltage regulator is shown in Fig. 3.8 Alternatively, a DC to DC converter may be used with greater efficiency ($\sim$90 %) to bring the supply within requirements.

3.6.1.2 Clocks

Like most digital equipments, microcontrollers require a synchronizing timing pulse provided by some form of clocking device. There are basically three common ways of implementing a timer in a microcontroller:

1. Internal clock
   The simplest option is to use the internal oscillator incorporated in the microcontroller itself. The required mode of oscillation is set while programming the controller.

Fig. 3.9 R-C network to be used as clocks for PIC® micro-controllers. Reproduced with permission from Microchip Technology Inc. (2008)

Fig. 3.10 Crystal oscillator circuit for PIC® controllers. Reproduced with permission from Microchip Technology Inc. (2008)

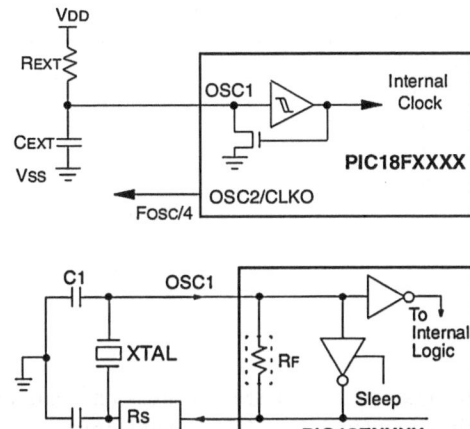

2. R-C network

A R-C circuit (Fig. 3.9) is connected to the microcontroller whereby the discharge of the connected capacitor provides the required frequency of operation of the microcontroller.

3. Crystal oscillator

A crystal oscillator is constituted by a flat thin slab of a quartz crystal, machined to precise dimensions whose opposite faces are metallized and attached to two connection pins. The crystal oscillator is electrically equivalent to a parallel L-C circuit that resonates at a precise resonant frequency (depending on the crystal's dimensions) and has a high Q-factor. When connected to a transistor and power supply, it produces a sinusoidal signal of highly precise frequency (Frerking 1996). The circuit for a crystal oscillator is given in Fig. 3.10.

The disadvantage of using the internal oscillator and the R-C network is that they are highly imprecise and this inaccuracy might lead to wrong calculations, hence unsatisfactory tracking. The best option remains the use of crystal oscillators that provides a stable clock signal whose accuracy is typically a few parts in a million (PPM).

3.6.2 Time Keeping

A Real-Time Clock (RTC) chip acts as an external clock that keeps track of the time while another digital device is busy performing an operation/instruction. The critical advantage of RTCs besides the fact that they are very accurate is that they can track time even if there is no power. It operates independently of the microcontroller attached to it.

An 8-pin package Dallas Semiconductor DS1307 chip (Table 3.3) is used for time-keeping. DS1307 provides full binary-coded decimal (BCD) clock/calendar plus

Table 3.3 Pin assignments for RTC DS1307 (Maxim Integrated Products 2008)

Pin number	Symbol	Description
1/2	X1/X2	32.768 KHz Crystal connection
3	V_{bat}	+3 V Power supply
4	GND	Ground
5	V_{cc}	+5 V Power supply
6	SQW/OUT	Square wave output
7	SCL	Serial data
8	SDA	Serial clock

56 bytes of NVSRAM. Address and data are transferred serially via a 2-wire, bi-directional bus. The clock/calendar provides second, minutes, hours, day, date month, and year information. The end of the month date is automatically adjusted for months with fewer than 31 days, including corrections for leap year (Maxim Integrated Products 2008). It is important to mention that the DS1307 has a built-in power sense circuit that detects power failures and automatically switches to battery supply. When the system is powered back on, the RTC IC feeds the system with the actual time and not the time before the power failure. This feature is of vital importance for a solar tracker since this allows the tracker to reorient itself if an extended power shortage was in effect.

3.6.3 Liquid Crystal Display

A liquid crystal display (LCD) is a pixilated output device capable of displaying characters and dot-based graphics. In operation, liquid crystal displays consist of two pieces of polarized glass sheets with perpendicular axes of polarity sandwiched between a layer of crystals in the liquid state. Depending on the current supplied, the liquid crystals twist and change the polarized plane, controlling the amount of light passing through the two polarizers (Chen 2011).

The LCD to be used is an alphanumeric type which is composed of linear segments. Segmented electrodes are suitable for simple alphanumeric displays and to display entire character sets or graphics, a dot-addressable matrix of electrodes is necessary. The LCD has two lines of 20 characters of 5×8 dots as depicted in Fig. 3.11 compatible with Densitron HD44780 with either 4 or 8 bits microcontroller interface. The interface circuit contains 14 pins, 3 out of which are reserved for supply voltage and dot contrast adjustment. Another 3 pins are used to control the operation of the LCD: RS (set/reset), R/W (read/write select), Enable (clock signal to initiate transfer of data). The remaining 8 lines are used for data transfer between the microcontroller and the LCD controller (see Table 3.4).

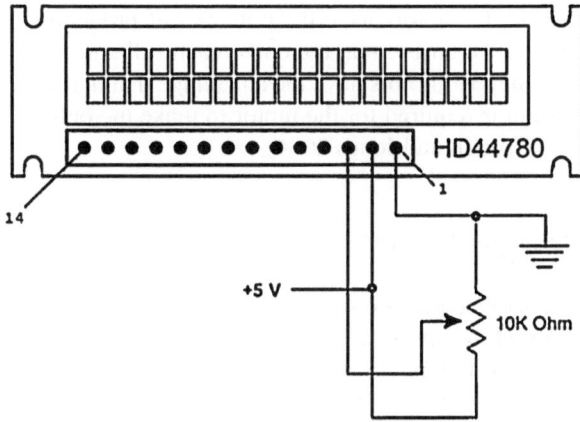

Fig. 3.11 Connections for supply power and dot contrast adjustment

Table 3.4 Pin assignments for Densitron compatible LCDs (Densitron Displays 2005)

Pin number	Symbol	Description
1	V_{ss}	Ground
2	V_{cc}	+5 V Power supply
3	V_e	Contrast control
4	RS	Set/Reset
		0 = instruction input
		1 = data input
5	R/W	Read/Write select
		0 = write to LCD
		1 = read LCD data
6	E	Enable signal for data transfer
7–14	DB0–DB7	Data bus line 0–7

3.6.4 Motor Control

3.6.4.1 Stepper Motor

A stepper motor is an electromagnetic actuator that accepts precisely timed pulse inputs and in response turns its output shaft clockwise or anticlockwise by a step angle or a few half-step angles depending on the sequence received and the type of device. Typically all the windings of a stepper motor are on the stator, the static outer cylindrical part, while the inner rotating cylinder called the rotor has a fixed number (24, 48, 72, 96, etc) of equilibrium positions. Due to their inductance, the windings don't instantly draw their full current and in fact may never reach full current at high stepping frequencies. The electromagnetic field produced by the coils is directly related to the amount of current they draw. The larger the electromagnetic field, the more torque the motors have the potential of producing. The solution to increasing the torque is to ensure that the coils reach full current draw during each step (Athani 1997). Stepper motors have the advantage that neither a positive sensor nor a feedback system is required for the motor to make the output response follow the input command (Condit and Jones 2004).

3.6.4.2 Types of Stepper Motors

There are three types of stepper motors:

Permanent magnet motors—have a large number of permanent magnets magnetized perpendicular to the axis and are arranged such that the polarity alternates from one segment to the next. Permanent magnet motors generally have large step angles and step at relatively low rates, but they can exhibit high torque and good damping characteristics (Fig. 3.12a).

Variable reluctance motors—are characterized as having multiple rotors (toothed blocks of some magnetically soft material) and a wound stator. They generally operate with small step angles at relatively high step rates, and have no detent torque[1] (Fig. 3.12b).

Hybrid stepper motors—have the better part of both the variable reluctance and permanent magnet stepper motors. The motor is multi-toothed like the variable-reluctance motor and contains an axially magnetized concentric magnet around its shaft. The teeth on the rotor provide an even better path which help guiding the magnetic flux to the preferred location in the air-gap. Overall, they have high torque, and they can operate at high stepping speeds (Fig. 3.12c).

3.6.4.3 Coil Excitation Types

The sub-variations of the different types of stepper motor types are determined by how the leads from each phase windings are brought outside of the motor.

Bipolar Stepper Motors

Bipolar motors are designed with two identical coils that are not electronically connected. The separate coils need to be driven in either direction for proper stepping to occur and for this, the polarity of the voltage across either coil must be reversed, so that current can flow in both directions, thus giving the name bipolar (Fig. 3.13a). This allows each stator pole to be magnetized to either north or south (Acarnely 2002). The adopted mechanism for reversing the voltage across one of the coils is called an H-Bridge, as it resembles the letter "H" (see Fig. 3.14). The current can be reversed through the coil by closing the appropriate switches—'AD' to flow one direction or 'BC' to flow the opposite way.

---

[1] Holding torque is the measured torque when the motor is stationary while detent torque is the torque when no current is flowing through the motor.

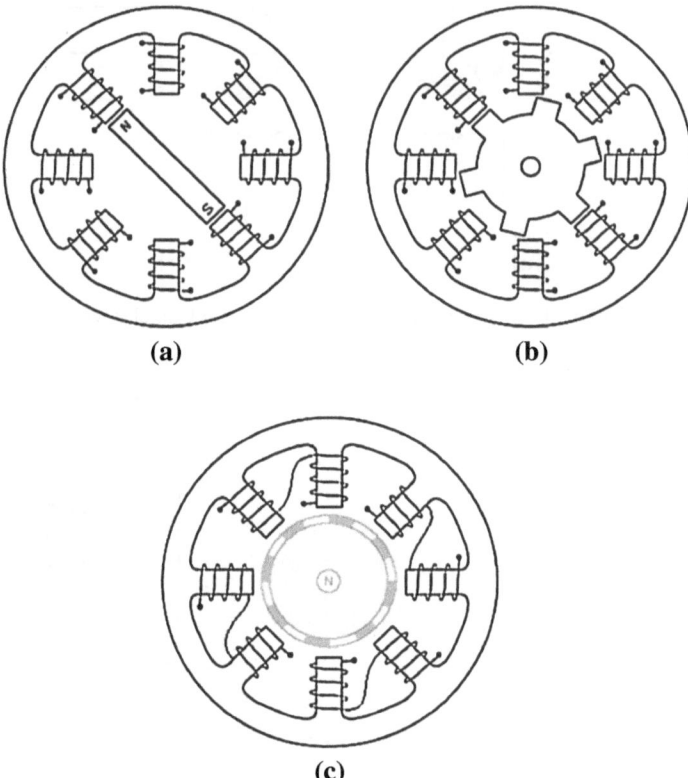

Fig. 3.12 Types of stepper motor. **a** Cross-section of a permanent magnet stepper motor. **b** Cross-section of a variable stepper motor. **c** Cross-section of a hybrid stepper motor

The motor itself is simple but the drive circuitry needed to reverse the polarity of each pair of motor poles is of high complexity.

Bipolar motors are reputed for their excellent size-to-torque ratio (torque is related to winding current)

Unipolar Stepper Motors
Unipolar stepper motors also have two coils, simple lengths of wound wire that are identical and are not electrically connected. Unipolar stepper motors are characterized by their centre-tapped windings—a wire coming out from the coil that is midway in length between its two terminals. Unipolar configuration allows current flow in half of the winding at one instant during operation (Acarnely 2002). Unipolar stepper motors, both permanent magnet and hybrid stepper motors with 5 or 6 wires are usually wired as shown in the schematic 3.13b.

A simple 1-of-'n' counter circuit can generate the proper stepper sequence for controlling unipolar motors. A common wiring scheme is to feed all the taps of the centre-tapped windings to the motor voltage. The driver circuit would then ground

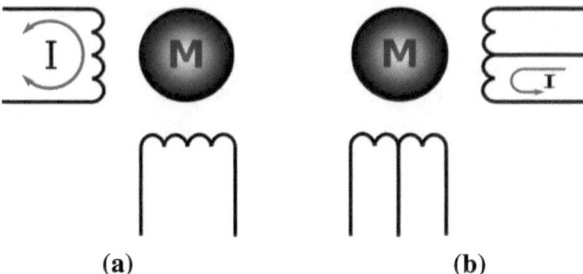

(a)   (b)

Fig. 3.13 Types of wiring in stepper motors. **a** Bipolar wiring. **b** Unipolar wiring

Fig. 3.14 H-Bridge circuit

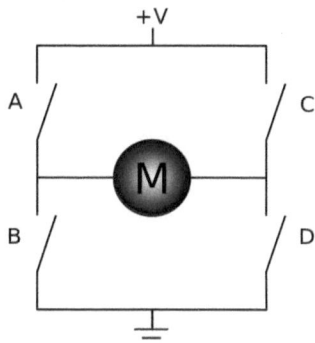

any winding to energize it such that the direction of the field provided by that winding is reversed.

Unipolar windings are thinner as compared to their bipolar counterpart. This implies that more wires are needed thereby increasing the windings resistance. This could cause an increased power loss via the windings potentially raising the temperature considerably.

3.6.4.4 Motor Selection for the Solar Tracker

Our interest lies in unipolar stepper motors, because it can be easily controlled with a microcontroller with little or no additional circuitry. The microcontroller will hence provide the required pulses to drive the motors. The motor that was chosen for the prototype was an EAD Rotary Hybrid Stepping Motor 1.8° LA23DGK-23 with a unipolar torque of 168 oz-in (1.21 Nm) and a bipolar torque of 210 oz-in (1.51 Nm). One reason for this choice is that EAD hybrid stepping motors are high precision bi-directional devices with position accuracy of ±3 % non-cumulative (ElectroCraft Inc. 2009). The specifications of this stepper motor are shown in Table 3.5.

Table 3.5 Specifications of EAD LA23DGK-23 stepper motor (ElectroCraft Inc. 2009)

Electrical Ratings

| | | Unipolar Connection | | | | | Bipolar Connection | | | | |
Model Number	Number of Leads	Phase Voltage (VDC)	Phase Current (amps)	Phase Resistance (ohms)	Phase Inductance (mH)	Holding Torque (oz-in)	Phase Voltage (VDC)	Phase Current (amps)	Phase Resistance (ohms)	Phase Inductance (mH)	Holding Torque (oz-in)
LA23DGK-23	6	6.00	1.76	3.40	8.4	168	8.50	1.25	6.80	33.4	210

Fig. 3.15 Typical torque provided by the EAD LA23DGK-23 motor as a function of stepping speed (ElectroCraft Inc. 2009). **a** Unipolar: Constant Voltage Drive (L/R), 2 Phase On. **b** Bipolar: 40 VDC Power Supply, 2A

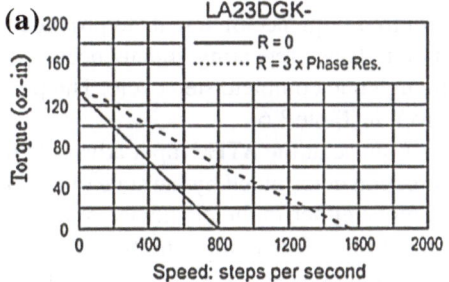

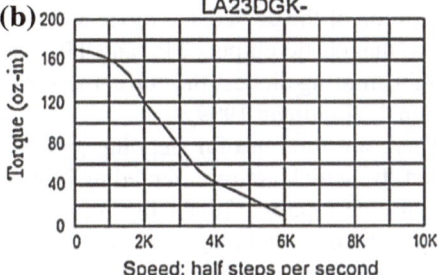

The pull-out torque-speed curves as tested by EAD engineers for both unipolar (Fig. 3.15a) and for bipolar operation (Fig. 3.15b) with a 40 V DC supply supplying a current of 2A is given by (Fig. 3.15b).

3.6.4.5 Stepper Motor Controller

A stepper motor must have an associated motor controller so that it is controlled efficiently, without missing steps. However, such controllers are very expensive and out of the allocated budget of this project. A compromise has to be found between accuracy, price and the StepGenie IC comes to the rescue. The chip emulates the

Table 3.6 Pin assignments for StepGenie IC (Laidman 1999)

Pin number	Symbol	Description
1	V_{dd}	+5 V Power supply
2/3	Mode a/b	Table 3.7
5–8	OUTD–OUTA	Output Line D–A
9/10	TP1/TP0	Diagnostic Test 1/0
11	STEP	Increment the step of motor
12	DIR	Direction of motion of motor
13	ENAB*	Enable signal for data transfer
14	GND	Ground

required sequence of signals to produce the correct stepping movement of a stepper motor. These repeating signals occur on four outputs lines and advance the motor though one complete step cycle. The pin-out diagram of the StepGenie controller is given in Table 3.6.

A pulse at the STEP input advances the motor by the step angle, i.e. 1.8°, in the direction determined by the state of the DIR signal whereby it specifies the direction to move either in the clockwise or anticlockwise direction. The ENAB signal must be held low for stepping motion to occur. As usual, V_{ss} and V_{dd} are the +5 V and 0 V inputs respectively. The four outputs (OUTA, OUTB, OUTC, OUTD) are the motor control output lines and the state of these outputs is determined by the current position of the device, the mode and the ENAB input. The StepGenie also provides two diagnostic outputs TP0 and TP1 which can be used to diagnose (or display by light emitting diodes) the current status of the device without loading input or output signals (Laidman 1999).

There are three useful stepping patterns, determined by MODE Input signals 'a' and 'b', which are represented in Table 3.7. After the last step in each sequence the sequence repeats. Advancing forward through the table causes the motor to turn one direction, stepping backward through the table causes the motor to turn in the opposite direction. The basic principle of stepper control is to reverse the direction of current through the two coils of a stepper motor, in sequence, in order to influence the rotor. Since there are two coils and two directions, that gives a possible 4-phase sequence as depicted in Table 3.7.

3.6.4.6   StepGenie versus PIC ®

Although stepping patterns can be programmed in a PIC® microcontroller, most PIC® microcontroller solutions are not satisfactory due to slow response/lag times between the step command and the actual change of the output signals. The effect can be system lag, or variation in response time which creates an uneven pulse train, resulting in motor jitter.

Table 3.7 Drive sequences for a unipolar stepper motor in wave-drive, high-torque and half step modes

Drive sequence	Step	Sequence				Description
		1	2	3	4	
Wave-Drive	1	ON	OFF	OFF	OFF	Only one phase is energized
	2	OFF	ON	OFF	OFF	at a time. Assures positional
	3	OFF	OFF	ON	OFF	accuracy regardless of any
	4	OFF	OFF	OFF	ON	winding imbalance in motor.
Wave Drive (Mode 'a' = ON & 'b' = OFF) → Single coil, 4-step pattern						
Hi-Torque	1	ON	OFF	OFF	ON	This sequence energizes two
	2	ON	ON	OFF	OFF	adjacent phases, offering an
	3	OFF	ON	ON	OFF	improved torque-speed
	4	OFF	OFF	ON	ON	product and greater torque.
High-Torque (Mode 'a' = ON & 'b' = ON) → Double coil, 4-step pattern						
Half-Step	1	ON	OFF	OFF	ON	Effectively doubles the step-
	2	ON	OFF	OFF	OFF	ping resolution of the motor,
	3	ON	ON	OFF	OFF	but the torque is not uniform
	4	OFF	ON	OFF	OFF	for each step. (Since we are
	5	OFF	ON	ON	OFF	effectively switching
	6	OFF	OFF	ON	OFF	between Wave-Drive and
	7	OFF	OFF	ON	ON	Hi-Torque with each step).
	8	OFF	OFF	OFF	ON	
Half-Step (Mode 'a' = OFF & 'b' = ON) → Mixed, 8-step pattern						

3.6.4.7 Motor Drive Transistors

Motor drive transistors operating under the supervision of the control circuits serve to shape the pulses that power the stepper motors. A power Metal Oxide Semiconductor Field-Effect Transistor (MOSFET) is a specific type of MOSFET designed to handle large amounts of power. Its main advantages are high commutation speed and good efficiency at low voltages (Sze 1998). Hexfets are N-Channel MOSFETs (Fig. 3.16) which have been especially tailored to minimize on-state resistance, provide superior switching performance, and withstand high energy pulse in the avalanche and commutation mode (Grant and Gowar 1989).

Fig. 3.16 Internal circuitry for HEXFETs

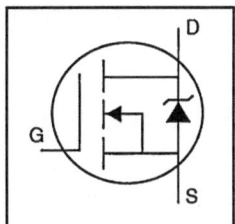

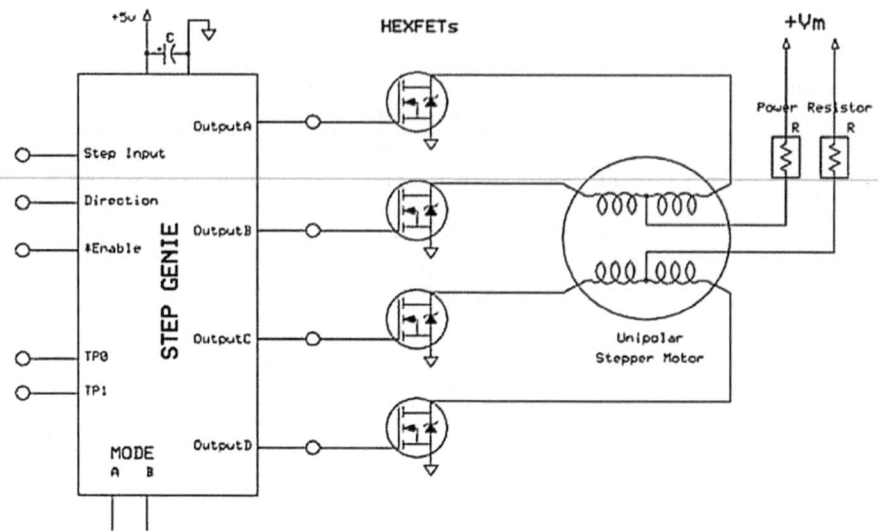

Fig. 3.17 Interfacing stepGenie with PIC® controller. Reproduced with permission from Laidman (1999)

In brief, it is the ideal component between a stepper motor and its digital counterparts, since it keeps the digital circuit 'safe' from the motor as it absorbs the surplus current that can damage the sensitive components. It is to be noted that protection diodes are not required if HEXFETs are used. In this project, IRLIZ24N HEXFETs have been used in between the two motors and the motor controller.

3.6.4.8 StepGenie Interfacing

The StepGenie IC must be interfaced with a microcontroller for operation, such that the latter sends the required STEP and DIR signal which is decoded by the StepGenie IC and moves the stepper motor in the required direction. Figure 3.17 shows the connections between any microcontroller and the stepper motor controller.

The four outputs are HEXFET compatible allowing the designer to tailor a power stage to the requirements of the application. A complete high-current unipolar application requires only five external components—four HEXFETs and a bypass electrolytic capacitor (10 μF 50 V) are recommended for general motor applications.

3.7 Circuit Implementation

Sections 3.6 and 3.6.4 have described the independent building blocks of the tracker circuit. The various subsystems were assembled and this consisted of interfacing the embedded systems, namely the PIC® microcontroller with the LCD, the real-time

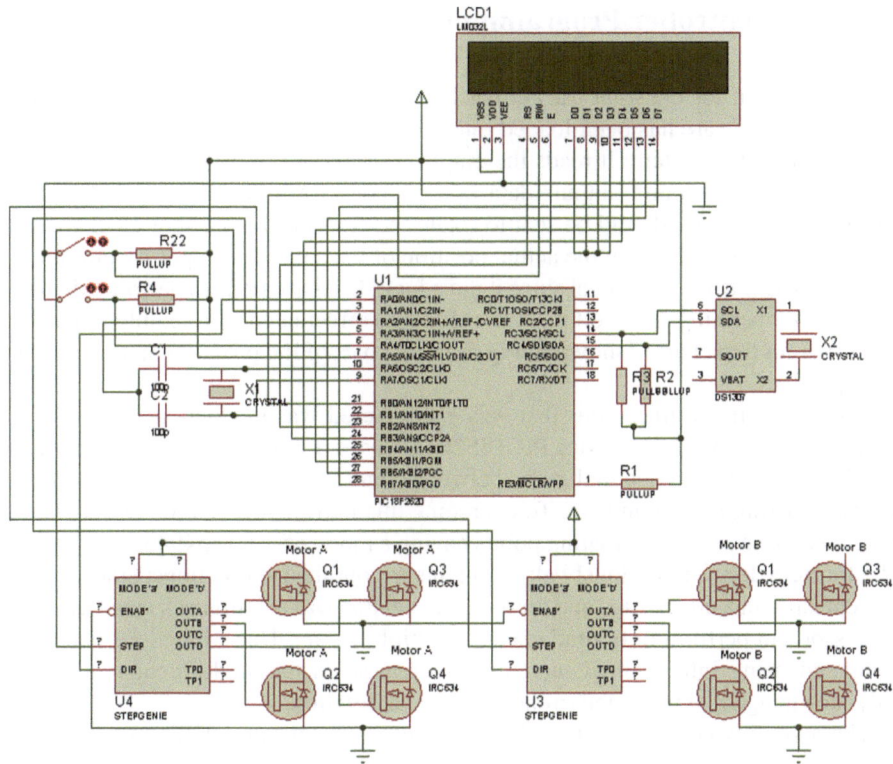

Fig. 3.18 Circuit diagram for solar tracker

clock and the motor controller. The complete schematic of the circuit is shown in Fig. 3.18. The design must ensure that the high current devices like the power resistors, HEXFETs or motors are secluded from the sensitive devices; fluctuations in current can cause inaccuracies in the measurement of time. In short, the analogue components are powered separately from the digital counterparts.

After the circuit design was completed, a computer prototyping software (FreePCB 1.358) was used to produce the printed circuit board (PCB) artwork output file in gerber format (*.gbr). This file was subsequently fed to a Bungard Computer Numerically Controlled (CNC) machine which milled the PCB out of a double-sided copper clad epoxy board. A complete step-by-step description of the whole procedure is given in Appendix B. Finally the components were soldered in place by hand using a temperature-controlled soldering iron. The microcontroller was not soldered directly on the PCB, but mounted on IC holders affixed to the circuit board. Adopting this approach ensured that no damage is done to the microcontroller and most importantly, the computer-on-a-chip could be removed for programming/debugging tasks.

3.8 Microcontroller Programming

After the circuit for the solar tracker has been realised, it is not functional until the appropriate software has been loaded into the microcontroller. The microcontroller is the brain of the system. It reads the output from the RTC, performs computation of the sun's position and outputs data in a readable form on the LCD. After that, the necessary signals should be sent to rotate the stepper motor either clockwise or anticlockwise depending on the actual position of the sun. The proper programming of the microcontroller is crucial since it must include instructions to perform all these tasks while operating. The coding of the chip was effectuated in assembly language. The program flowchart is shown in Fig. 3.19 and part of the program listing is given in Appendix C.

One of the main difficulties that was encountered was to calculate the position of the sun with high precision; PIC18F2682® uses a byte to represent a number and this number has a decimal range between 0 and 255 but calculations need to be done in floating point numbers. To overcome this barrier, 4 bytes (0–4294967295) were merged to represent floating points having a range of $\pm 1.17549435082 \times 10^{38}$. The formula derived previously also contains trigonometric functions: sine, cosine, tangent, arc-sine, arc-cosine, and arc-tangent and this was a bottleneck since a microprocessor can perform only fundamental multiplications. These functions had to be computed using infinitesimal increment/decrement in the angle recursively since only the change had to be determined with each iteration.

The assembly codes were tested by first converting it into HEX codes and then burning it into the chip using a PIC® microcontroller compatible flasher. The functionality of the tracker dictated the success or failure of the program. If the output from the chip was not as expected, then, the codes were modified and the PIC® microcontroller was flashed again until a satisfactory result was obtained, using a modify-test recursive approach.

3.9 Prototype Implementation and Realisation

A prototype had to be implemented with the following characteristics;

Rigidity—Because the hardware is to be placed in the open air, it should be resistant to fierce weather conditions. Concrete, stainless steel or plated metals are possible choices. Unfortunately wood or iron is not an alternative although they are cheap as wood will rot once in contact with moisture just like iron would rust.

Viewing angle—The prototype should be able to locate any specific point in space. Thus, two motors had to be placed perpendicular to each other, one will make the whole system rotate in the $x-y$ plane and another independent motor will make part of the system rotate in the $r-z$ plane.

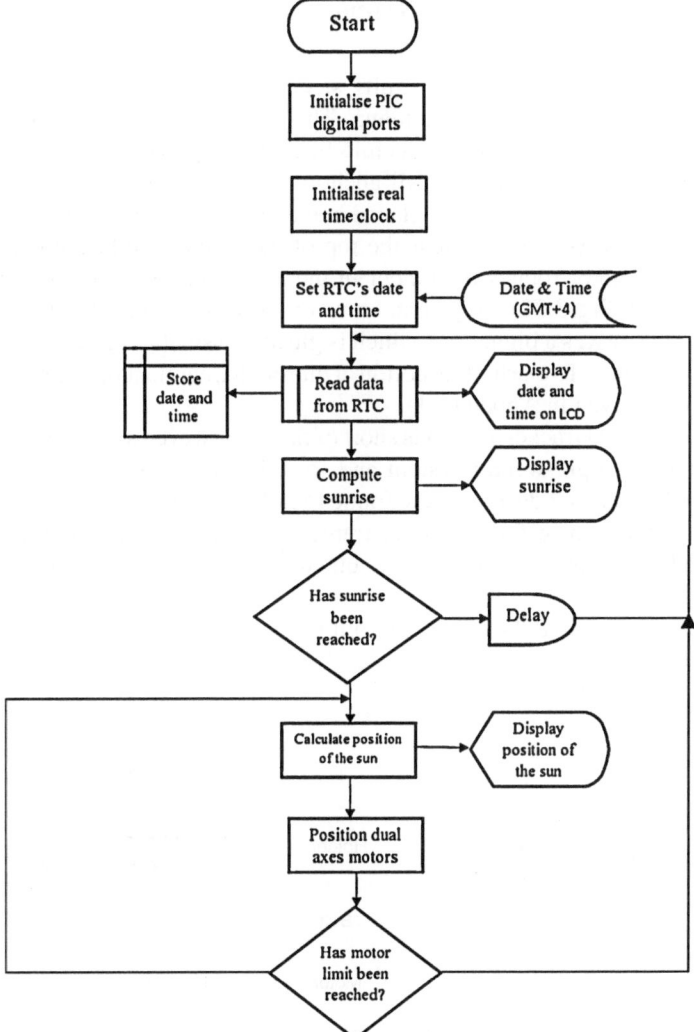

Fig. 3.19 Microcontroller flowchart for solar tracking

Ease of manufacture—The material required to build the system should be readily available. In addition, the tools required should be acquirable in a reasonably equipped workshop.

Portability—The final system should be light-weight and portable. Being a proof of concept, the system should face challenging experiments so as to validate future systems of similar kinds.

Cost—The bits and pieces used to build the prototype should not be expensive. The low-cost advantage will certainly make the solar tracker a tempting system.

3.9.1 Mode of Operation of System

The idea that comes to mind when fully steerable tracking systems is mentioned is undoubtedly satellite tracking antenna. This fully steerable tracking antenna is a simple yet robust system. The principle behind its operation is rather simple to understand. A concrete pillar is well-anchored in the ground and just above the ground is mounted a stepper motor placed in the axial direction of the pillar to rotate the machine in the $x-y$ plane. Near the top of the frame, another motor placed at $90°$ to the first motor rotates the holder in the $r-z$ plane. Both motors are located inside the frame (Fig. 3.20). To gain in terms of torque, two sets of pulleys are used; a smaller pulley drives a bigger one which is glued to a smaller pulley which in turns drive a larger pulley. To clarify this notion, a thorough treatment on the magnification of the torque is given in Appendix D.

The conventional tracker system is short of flaws. However we are more interested in a portable and light-weight system and for this reason, we have to discard the conventional frame and design a new frame to hold the solar concentrator.

The preferred choice for the novel frame's material was aluminium, which is resistant to bad weather conditions, immune to chemical substances like water and light-weight.

The unexampled modular system is split into two distinct parts: the base and the top platform.

Fig. 3.20 Conventional tracker

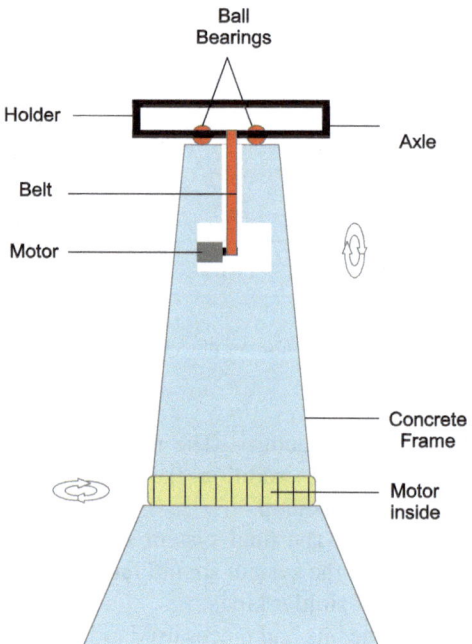

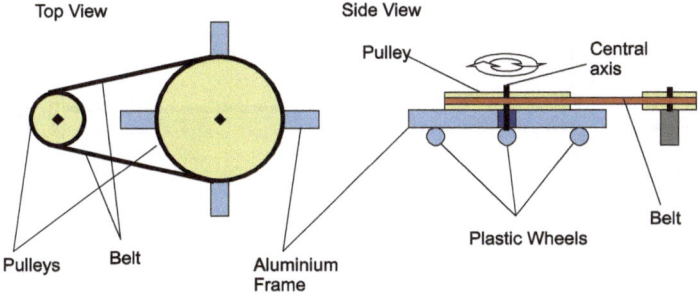

Fig. 3.21 Base of solar tracker

The base is cross-shaped to ensure system's stability. It is supported by plastic wheels, and an axle protruding at its centre permits the system to rotate about this axis. The wheels minimise friction while at the same time, relieves the stress on the central axle. Fitted to the axle is a pulley which, with the help of a suitable belt, connects the system to a smaller pulley which is glued to one of the two motors as depicted in Fig. 3.21. Operating this motor rotates the system as a whole about the central axis (x–y plane).

The base is linked to the top platform with the help of an aluminium tube. It is noteworthy to mention that the frame has a U-shape resting on one of its side. This frame is chosen instead of an I-shape frame because it allows a motor to be mounted on its side as shown in Fig. 3.22. Both the U-shape and the I-shape frames perform similarly while operating.

The final component of the prototype is the top platform which is designed to hoist a receiver. Resting on it is an axle, supported by a pair of ball-bearings which ensures the proper operation by the diminution of the forces of friction. A flat plate is soldered to the axle such that different types of receivers can be installed for a variety of applications (e.g photovoltaic modules, solar thermal tubes). A pulley is

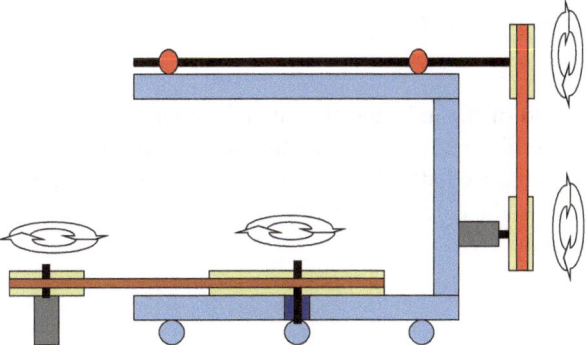

Fig. 3.22 Skeleton of solar tracker

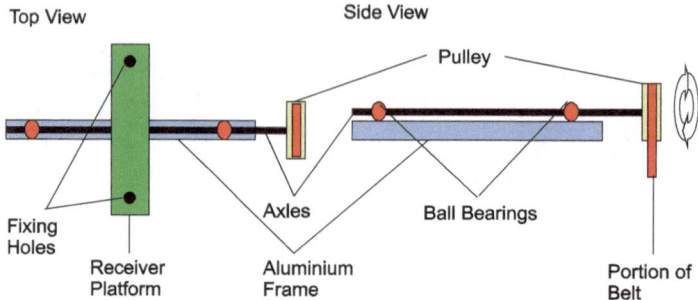

Fig. 3.23 Top skeleton of solar tracker

Fig. 3.24 Prototype of solar tracker

anchored to one end of the axle and the latter is connected to a motor fixed to the frame by a belt (see Fig. 3.23), permitting the machine to move in the $r-z$ plane.

The system was realised after the design stage was over. Finally, the mechanical system was linked to the electronic circuit and Fig. 3.24 shows the implementation of the tracker system.

Table 3.8 Actual cost of tracking solar concentrator

Equipment	Unit price ($)	Debit ($)	Total Cost ($)
Circuit			
PIC18F2682® Microcontroller	10.32	10.32	10.32
DIP sockets—28pins	0.44	0.44	10.76
32.768 Crystal oscillator	0.42	0.42	11.18
HD44780 LCD	15.00	15.00	26.18
7805 Voltage regulator	0.80	0.80	26.98
DS1307 Real-time clock IC	4.45	4.45	31.43
DIP sockets—14pins	0.24	0.24	31.67
StepGenie IC (X4)	6.00	24.00	55.67
DIP Sockets—8pins (X4)	0.19	0.76	56.43
Hexfets (X32)	1.72	55.04	111.47
Power resistor (X2)	0.63	1.26	112.73
Assorted resistors	0.50	0.50	113.23
Assorted capacitors	0.50	0.50	113.73
Double-sided copper PCB	5.08	5.08	118.81
Frame			
Stepper motor (X2)	40.00	80.00	198.81
Aluminum tube	16.32	16.32	215.13
Pulleys and bearings (X4)	8.50	34.00	249.13
Timing belts (X4)	4.00	16.00	265.13
Assorted screws and bolts	2.00	2.00	267.13
Cement (X3)	3.66	10.98	278.11
Limit switches (X2)	0.24	0.48	278.59
		Subtotal	278.59
		Shipping	28.96
		Total cost	307.55

3.10 Solar Tracker Cost

Once the solar tracker was implemented, a simple account was made illustrating in Table 3.8 the actual cost of the system.

It is important to observe that the electronics circuitry is universal in the sense that *any* solar tracker can utilize it. The cost of producing such a general circuit was only $118.81 excluding shipping charges.

Our tailor-made frame was priced at $159.78 and if the conventional frame was to be built then, the price would increase astronomically since the site has to be prepared, lots of raw material have to be used and huge manpower is needed to accomplish the frame. Even with the conventional frame, the brain of our solar tracker can still be used to steer the system.

3.11 Chapter Summary

The various types of solar trackers are reviewed in this chapter along with their merits and demerits. This taxonomy has proved to be useful while designing the final prototype system because it gives a clear-cut idea of what we want to incorporate in our solar tracker. It has been shown that in terms of the relative power output, a dual-axes tracker is the most efficient system available.

The step-by-step construction of a novel dual-axes solar tracker, that points directly towards the sun thanks to an integrated sun tracking mechanism with two degrees of rotational freedom, is presented in this chapter. Each stage of the design, with explicit explanation of all the components, and realisation of the solar tracker is detailed. The electro-mechanical control system is based on a precisely-timed micro-controller circuit that first computes the altitude and azimuth of the sun in real-time and then drives a pair of stepper motors that steers the system towards it. The system will track the sun throughout the day and return to its default position for night-time stowing.

The whole set-up can be constructed in about 6 months at a record price of \$118.81 for the electronics circuitry that *any* solar tracker can utilize and \$159.78 for a tailor-made prototype frame. Future generations of the tracker would be based on the same core with more robustness and prolonged usage in the open, at the mercy of rainy conditions, temperature variations and windy weather.

References

Acarnely P (2002) Stepping motors: a guide to theory and practice, 4th edn. The Institution of Engineering and Technology, Herts

Athani V (1997) Stepper motors: fundamentals. Applications and design, New Age International Limited, New Delhi

Barsoum N (2011) Fabrication of dual-axis solar tracking controller project. Sci Res ICA 2(2):57–68

British Computer Society (2002) A glossary of computing terms, 10th edn. Pearson Education Limited, Essex

Chen R (2011) Liquid crystal displays: fundamental physics and technology. Wiley, New Jersey

Condit R, Jones D (2004) Stepping motors fundamentals. Microchip Application AN907

Cotfras D, Cotfras P, Kaplanis S, Ursutui D, Samoila C (2008) Sun tracker system versus fixed system. University of Brasov http://www.but.unitbv.ro/bu2008/BULETIN%20III%20PDF/Cotfras-rez.pdf. Accessed 16 Jan 2009

Densitron Displays (2005) LM2053 datasheet. Datasheet http://www.densitron.com/GetPdf.aspx?nDisplayID=523. Accessed 20 Nov 2012

ElectroCraft Inc (2009) Ead rotary hybrid stepping motor - la23dgk-23 datasheet. Datasheet http://www.electrocraft.com/files/ead_step.pdf. Accessed 28 Feb 2009

Frerking M (1996) Fifty years of progress in quartz crystal frequency standards. In: Proceedings of the 1996 IEEE international frequency control symposium, IEEE, pp 33–46

Grant D, Gowar J (1989) Power MOSFETS: theory and applications. Wiley, New Jersey

Huang Y, Kuo T, Chen C, Chang C, Wu P, Wu T (2009) The design and implementation of a solar tracking generating power system. Eng Lett 17(4):1–5

Laidman R (1999) Stepgenie ic. Datasheet http://www.stepgenie.com/StepGenieSpec.pdf. Accessed 16 Dec 2009

Maxim Integrated Products (2008) Ds1307 rtc datasheet. Datasheet http://datasheets. maximintegrated.com/en/ds/DS1307.pdf. Accessed 07 Feb 2008

Microchip Technology Inc (2008) PIC18F2682/2685/4682/4685 datasheet. Datasheet ww1.microchip.com/downloads/en/DeviceDoc/39761b.pdf. Accessed 08 Oct 2008

Mousazadeh H, Keyhani A, Javadi A, Mobli H, Abrinia K, Sharifi A (2009) A review of principle and sun-tracking methods for maximizing solar systems output. Renew Sustain Energy Rev 13(8):1800–1818

Sze S (1998) Modern semiconductor device physics. Wiley, New Jersey

Chapter 4
Solar Concentrators

Abstract Concentrating solar technologies are in different stages of development; most of them have passed the testing and power production (on a small scale) phases and are being commercialised. Yielding the most power per area among all the solar concentrators is the parabolic dish and the latter was selected as our low-budget prototype of choice. The steps towards the final parabolic dish concentrator made with fibreglass have been enumerated in the pages of this chapter. To convert the parabolic dish to a solar parabolic concentrator, its surface has to be lined up with a reflective material so as to focus energy optimally. The different types of reflective materials have also been discussed with the pros and cons pointed out for each type. The choice of the solar concentrator along with the building materials, fibreglass and chrome vinyl reflector, were also fully justified in accordance with the ease of production and tight budget.

In this chapter, the different types solar concentrators from large-scale commercial systems to simple applications will be presented. This review will be used as a guide in our own endeavour to plan, design and implement a concentrator for use in conjunction with the solar tracker.

At the focus of a solar concentrator, a high solar irradiance prevails and the corresponding solar energy can be fed to a vacuum-tube producing thermal heat which can be used directly for thermal heating or to produce superheated steam that can be used to drive a generator for the production of electricity. It can also be used directly to produce electricity using highly efficient photovoltaic (concentrated PV) cells. The two mentioned system can also be combined: solar cells are primarily used as the receiver and a fluid is passed to cool off the cells, hence keeping the cells at optimal production temperatures. Last but not least, a Stirling engine, placed at the focus, can convert an externally-applied temperature differential into electricity by heating one end of the engine, while the other end is cooled by a closed-loop cooling system (Trieb et al. 1997).

Z. Jagoo, *Tracking Solar Concentrators*, SpringerBriefs in Energy,   49
DOI: 10.1007/978-94-007-6104-9_4, © The Author(s) 2013

4.1 Linear Concentrators

Among the several distinct types of solar power collectors in use are the linear concentrators which comprise of the parabolic trough and the Fresnel reflectors. Both will be discussed in this section.

4.1.1 Parabolic Trough

The parabolic trough captures the sun's radiation with mirrors-like materials that reflect and focus the sunlight onto a linear receiver tube which is positioned along the focal line (Fig. 4.1). The tube can be a fluid carrier for heating purposes or steam for the generation of electricity (Fernandez-Garcia et al. 2010).

Solar parabolic trough systems are the most proven and commercially tested solar concentrating power technology, primarily because of the nine large commercial-scale solar power plants that are operating in the California Mojave Desert (354 MW) (Price et al. 2002). Another commercial company, Nevada Solar One, uses linear

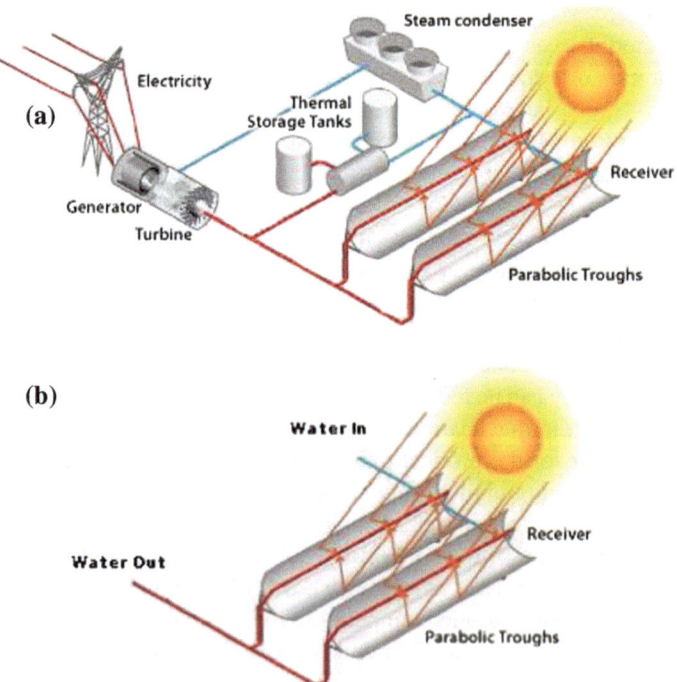

Fig. 4.1 Linear trough solar concentrator. **a** Linear trough concentrator for the generation of electricity (U.S. Department of Energy 2012b). **b** Trough concentrator for thermal heating

parabolic troughs as its core technology to mass-producing a nominal capacity of 64 MW and maximum capacity of 75 MW with an electricity production of 134 million kWh per year, as of June 2007. The power plant uses 760 parabolic troughs (using more than 180,000 mirrors) that concentrate the sun's rays onto thermos tubes running laterally through the troughs and containing a heat transfer fluid that produces electricity by driving turbines (Cohen 2006).

Because of their shape, parabolic troughs are limited to single-axis tracking mechanisms. For this reason, they lose part of the energy for not being to track the sun with high-precision and hence do not produce fluid temperature as high as some other solar concentrating technologies rendering its efficiency lower.

4.1.2 Fresnel Reflector

In this set-up, several Fresnel mirrors are used to reflect sunlight onto a linear receiver tube fixed in space above these mirrors (Fig. 4.2). Just as for the linear parabolic trough, the receiver can also be either used for the thermal heat production or generation of electricity or both (Abbas et al. 2012).

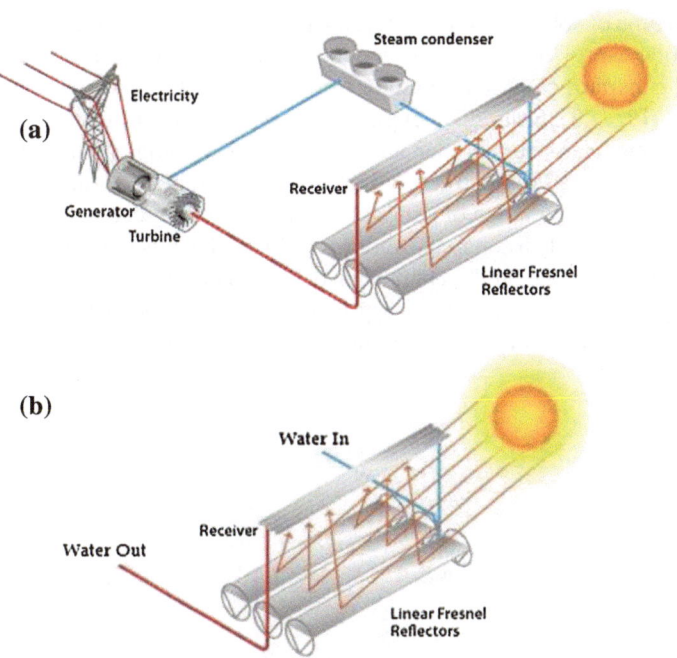

Fig. 4.2 Linear Fresnel solar concentrators. **a** Linear Fresnel mirrors for the generation of electricity using turbines (U.S. Department of Energy 2012). **b** Fresnel concentrators for heating fluid

Compared to linear trough systems, the compact linear Fresnel reflector system reduces costs by replacing heat-curved reflectors with standard flat Fresnel mirrors to concentrate the solar energy. Moreover, the Fresnel mirrors have typically a very short focal length and a large aperture which reduces the overall size of the reflector. Fresnel reflectors do not have a rich history like parabolic troughs but they started to gain momentum as from the twentieth century. For example in Bakersfield, USA, is located Ausras Kimberlina Solar Thermal Energy Plant and the collector lines generates up to 25 MW of thermal energy to drive a steam turbine and at full output, the Kimberlina facility produces enough solar steam to generate 5 MW of electricity. In addition to Kimberlina, Ausra is developing a 177 MW solar thermal power plant for Pacific Gas and Electric Company (PG&E) in Carrizo Plains (USA) (Fishman 2008).

However the sun is tracked in a single-axis only because the physical dimensions of the system causes a constraint and just as its counterpart, the parabolic trough, it suffers from not being able to track the sun with a good accuracy thereby a reduction in its efficiency.

4.2 Parabolic Dish

The second type of solar concentrators is the parabolic-shaped concentrator that concentrates solar energy onto a receiver mounted at the focal point. A power conversion unit which generates electricity directly from the concentrated solar energy is fitted at the focus (Fig. 4.3). Solar dish–engine systems convert sunlight into electricity at very high efficiencies—higher than any other solar technology (Wu et al. 2010). The dish must be mounted on a structure that tracks the Sun from sunrise to sunset to reflect the highest percentage of sunlight possible onto the receiver.

In theory, a single dish can make about 10,000 peak watts of heat and 3,500 peak watts of electricity in the Sunbelt and when deployed in large, utility-scale fields, this could make a difference to how energy is generated (Chandler and Ahrens 2008). One of the major advantage of parabolic dishes is that they make use of dual-axes

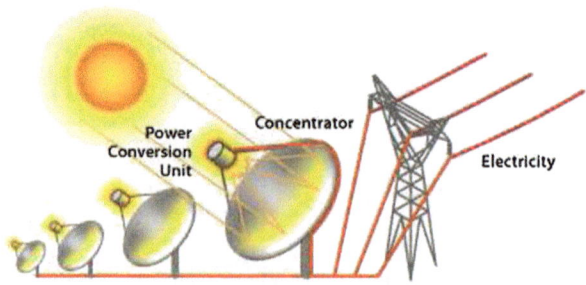

Fig. 4.3 Parabolic Dishes for generating electricity (U.S. Department of Energy 2012a)

tracking mechanisms which causes an enhancement in the energy production making them the most efficient solar concentrator in existence.

As of now, there has been no commercial outbreak of the parabolic solar concentrator because the implementation costs are exorbitant as accurate parabolic dishes are scarce and difficult to produce implying the expensiveness.

4.3 Power Tower Systems

A power tower makes use of numerous large, sun-tracking flat mirrors aligned together to focus sunlight on a receiver atop a tower (Fig. 4.4). In most of the towers, there is a heat transfer fluid in the receiver that is heated by concentrating sunlight which is used to generate steam. The plant pipes the pressurised steam from each thermal receiver and aggregates it at a turbine which powers a standard generator to produce electricity. The steam then reverts back to water through cooling, and the process repeats itself (Rabl 1976).

As with parabolic concentrators, power towers also make use of systems that are programmed to track the sun from dawn to dusk in a two-axis geometry. The mirrors always point in the direction of the sun which increases the efficiency drastically. One advantage of this system is the reliability of the system in all wind conditions since low wind profile tracking mirrors are used. Solar power towers offer large-scale, distributed solutions to the global energy needs, in particular for peaking power as exemplified by PS10 and PS20. Located just outside of Seville in Southern Spain is PS10, the first of two solar tower and heliostat field technology power plants that generates 11 MW while the second, PS20 generates a much higher 20 MW using 1255 mirrors, the two of which together would displace a total of 54,000 tons of carbon dioxide each year. By 2013, a conglomerate of eight power stations, the Sanlucar la

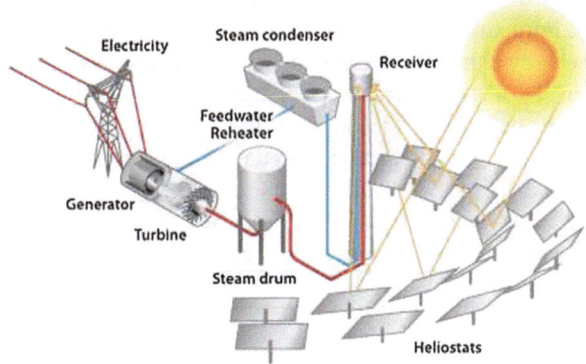

Fig. 4.4 Solar power tower (U.S. Department of Energy 2012d)

Mayor Solar Platform, will be able to produce a staggering 302 MW; enough energy to power 181,200 homes (Mancini et al. 1997).

Because an array of mirrors has to be used for optimal efficiency, this set-up is rather cumbersome compared to the other systems. Venturing in the periphery of a tower power system is hazardous because electromagnetic energy is concentrated from a large distance to the solar power tower.

4.4 Summary of Solar Concentrators

Like all solar technologies, solar concentrators are fuelled by sunshine and do not release greenhouse gases. In this section, we sum-up the different types of solar concentrating technologies that are available in Table 4.1. We should note that most of the concentrating solar power described in the previous sections cannot operate as a stand-alone system since the energy output would be too low and thus a large number of collectors in parallel rows, called a concentrating collector field, must be used.

4.5 Reflective Materials

Concentrators often use a reflective material to focus sun rays to a sharp spot. A critical task in developing a solar concentrator is to identify the best possible reflector materials. The perfect material would of course be one that provides high optical reflectance, UV ray resistant, durable in a variety of abusive environmental conditions, easily attachable to a substrate, and last but not least, economical. The materials currently being used range from common kitchen aluminium foil to complex films developed by major corporations:

(a) **Aluminium Foil**—The simplest reflective material is the common kitchen aluminium foil. Pure aluminium develops a protective coating of aluminium oxide immediately upon exposure to oxygen. This protective layer however decreases its reflectivity by a small percentage. Aluminium foil (thickness <0.2 mm) is not designed to be used in the outdoor, thence to last for extended periods of time in the open, it must be coated with a protective epoxy coating.

Table 4.1 Summary of the different type of concentrators

Type of concentrator	Focus type	Reflective material	Tracking
Parabolic dish	Point	Mirror	Two-axes
Power tower	Point	Mirror	Two-axes
Fresnel reflector	Line	Lens	Single-axes
Parabolic trough	Line	Mirror	Single-axes

(b) **Silvered Mirror**—Smooth glass (substrate) is silvered (reflective coating) from the backside and sealed with an oxidation protective layer to produce the common mirror with reflectivities higher than 85 %. Although glass mirrors have a high percentage of reflectivity, their applications are restricted mainly to flat profiles like for example in power towers or parabolic reflectors. Curved glass mirrors, traditionally require precise grinding and milling operations that render them extremely costly. Flabeg, the manufacturer of duraGLARE films, claims to have yielded mirrors with a reflectivity of minimum 94 % which is being used with success in LUZ-LS3 collectors since 1991 at Harper Lake in Louisville, Colorado (FLABEG Holding GmbH 2012).

(c) **Alanod Front Surface Aluminised Reflector**—The Alanod front surface aluminised reflector is a thin film comprising of entirely aluminium that has a total reflectivity of 95 %. Samples, coated with a polymeric chemical to protect the alumina layer, have survived outdoor exposure of more than 3 years under the SolarPACES project in Koln, Germany (Harrison 2001).

(d) **ReflecTech Mirror Film**—ReflecTech Mirror Film is a polymer-based film for concentrating sunlight in solar energy arrays. The film has an overall reflectivity of 94 % and is immune to water and UV radiation. Several ReflecTech Mirror Film facets have been in use at Kramer Junction SEGS VI for six years and with very little loss of reflectance (ReflecTech, Inc. 2009).

4.6 Realisation of the Solar Concentrator

The parabolic dish was chosen to be used in conjunction with the solar tracker designed because it is the most efficient concentrator in terms of size and power output while being compatible with our two-axes tracking system. In addition, a scalable lab prototype can be implemented with locally available raw materials. In this section, the different steps undertaken in the design of the parabolic dish concentrator will be elaborated. The parabolic dish is in fact a paraboloid of revolution, a surface obtained by revolving part of a parabola about its axis of symmetry. The parabola may be represented on a 2D flat surface by the equation $y = ax^2$, with the y-axis being the axis of symmetry of the parabola. In operation, a parabolic mirror surface receives light travelling parallel to the z-axis and focusses it at the focal point as shown in Fig. 4.5. The focal point is situated on the y-axis at point $(0, \frac{1}{4a})$.

To derive the relationship between the focal length and the constant a, we have to resort to a line which is parallel to the x-axis but beneath the parabola by a distance of f (Fig. 4.6). The property of the line $y = -f$, called the linear directrix, is that any point $P(x, y)$ on the parabola will be equidistant from both the focus and the line.

The length FP is given by

$$FP = \sqrt{x^2 + (f - y)^2} \tag{4.1}$$

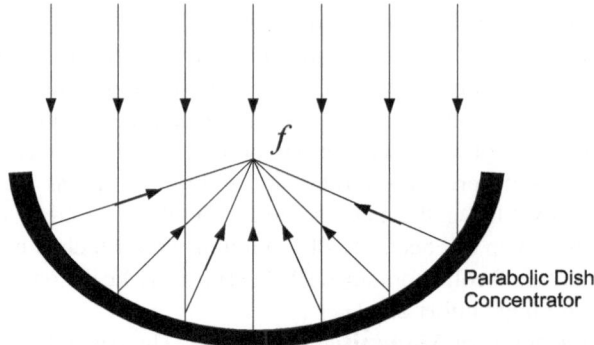

Fig. 4.5   Diagram showing parallel rays being focussed at f

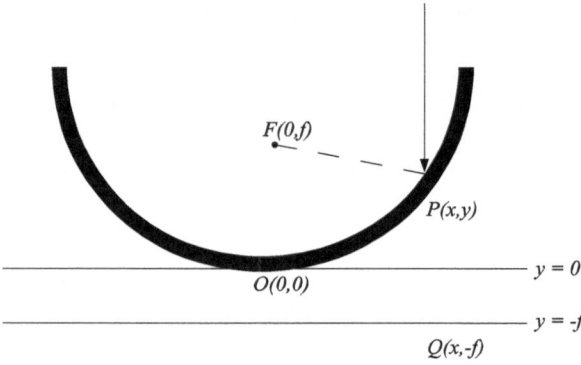

Fig. 4.6   Diagram showing the linear directrix, a general point and the focus

and QP which is the distance perpendicular to the linear directrix towards P.

$$QP = f + y. \qquad (4.2)$$

Since FP and QP are equal in distance, we can equate the 2 above relationships.

$$QP = FP \qquad (4.3)$$

$$f + y = \sqrt{x^2 + (f - y)^2} \qquad (4.4)$$

$$f^2 + 2fy + y^2 = x^2 + (f - y)^2 \qquad (4.5)$$

$$f^2 + 2fy + y^2 = x^2 + f^2 - 2fy + y^2 \qquad (4.6)$$

$$x^2 = 4fy \qquad (4.7)$$

$$x^2 = 4f(ax^2) \qquad (4.8)$$

$$f = \frac{1}{4a} \qquad (4.9)$$

The focal distance measured from the vertex of the paraboloid is therefore $f = \frac{1}{4a}$ as shown above.

4.6.1 Construction of the Parabola

We now have the simple mathematical background to calculate all the parameters of a parabola. Given the tools at our disposal, our objective will be to construct a parabolic dish of 90 cm diameter, that will be light enough to be steered by the tracker mechanism. We found it practical to opt for a focal length of 50 cm for our design (Table 4.2).

With these dimensions, the paraboloid would have a footprint of $0.64\,\mathrm{m}^2$. Fabrication of the low-cost paraboloid was effectuated by first drawing the profile of the parabola on a flat sheet of plywood. Then the parabola shape was cut out using a jigsaw and the outer parabola affixed to a pole which when rotated through 360° would define the actual paraboloid of revolution. Photographs depicting the various stages of the construction are displayed in Fig. 4.7a, b. We should reiterate that this paraboloid could be made with a much greater accuracy in a professional workshop but our aim is to construct a very low cost parabolic dish with basic tools.

The next step consisted in using this wooden outer-parabola to make a cement paraboloidal surface that would be used as a mould for casting a fibreglass parabolic dish. Rubble was piled and cement was added on top to save on costs instead of using cement only. The wooden form was inserted in the middle and rotated until the cement was spread evenly in the form of a parabolic dome (Fig. 4.8). The cement was left to cure for a few days in order to obtain a rigid solid surface.

The dome was polished[1] with fine sand-paper to even out inhomogeneities. At this stage, the cement dome is ready for application of fibreglass layer. The advantages of this rustic method are that firstly, it is extremely cheap and secondly, mass production of the dome is a viable option.

Table 4.2 Coordinates used for design of parabola with focal point 50 cm

x/cm	2.00	4.00	6.00	8.00	10.00	12.00	14.00	16.00	18.00
y/cm	0.02	0.08	0.18	0.32	0.50	0.72	0.98	1.28	1.62
x/cm	20.00	22.00	24.00	26.00	28.00	30.00	32.00	34.00	36.00
y/cm	2.00	2.42	2.88	3.38	3.92	4.5	5.12	5.78	6.48
x/cm	38.00	40.00	42.00	44.00	46.00	48.00	50.00		
y/cm	7.22	8.00	8.82	9.68	10.58	11.52	12.50		

---

[1] Note: Protective mask should be worn when polishing, so that the dust particles do not cause any damage to the respiratory system.

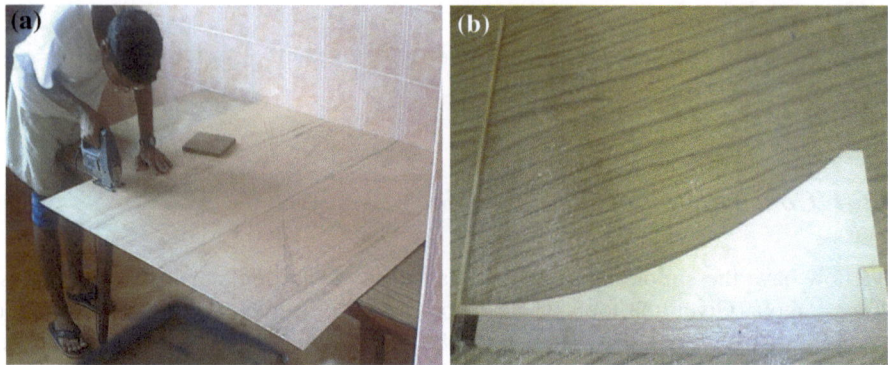

Fig. 4.7 Initial steps to make a parabolic dish. **a** Cutting wood with jigsaw. **b** Outer parabola with a pole

Fig. 4.8 Making of parabolic dome. **a** Rubble works. **b** Laying cement on rubbles

4.6.2 Fibreglass Layout

Prior to the moulding process, a thin film of silicone paste is applied to the parabolic dome as shown in Fig. 4.9a. This was done to facilitate the removal of the fibreglass after it had hardened. Next, alternate layers of fibreglass mat (C.S Glassmat EM 300) and resin layers (Orthophatic 901 PA) were coated on the concrete paraboloidal surface (Fig. 4.9b). This process was repeated until a satisfactory thickness was obtained. Fibreglass was the chosen material because of the following characteristics:

- cheap,
- impermeability,
- no previous know-how is needed (fast learning curve),
- can be moulded to any shape and size.

After leaving to dry for about a day, the fibreglass was removed, the excess fibreglass near the edges was trimmed, and ultimately, polished.

Fig. 4.9 Fibreglass layout. **a** Paraboloid after application of silicone paste. **b** Fibreglass being laid

4.6.3 Reflective Material Layout

High-reflective materials like Alanod front surface aluminised reflector or ReflecTech mirror films cannot be used when designing a limited-budget solution. Aluminium foil was not the best of choices because it is very thin and non-resistant. Instead, a chrome polymer reflector with an adhesive back for easy application was used. This impervious material is highly durable since it was made to withstand different weather conditions. 82 % of the incident sunlight, taking account soiling losses and cavity losses due to optical aberration, is reflected onto the focus of the parabolic dish.

The reflective vinyl was pasted to the fibreglass. The procedure consisted of cutting the reflective material into stripes and glue them as shown in Fig. 4.10a. The finished dish with a diameter of 90 cm and a focal length of 50 cm is shown in Fig. 4.10b.

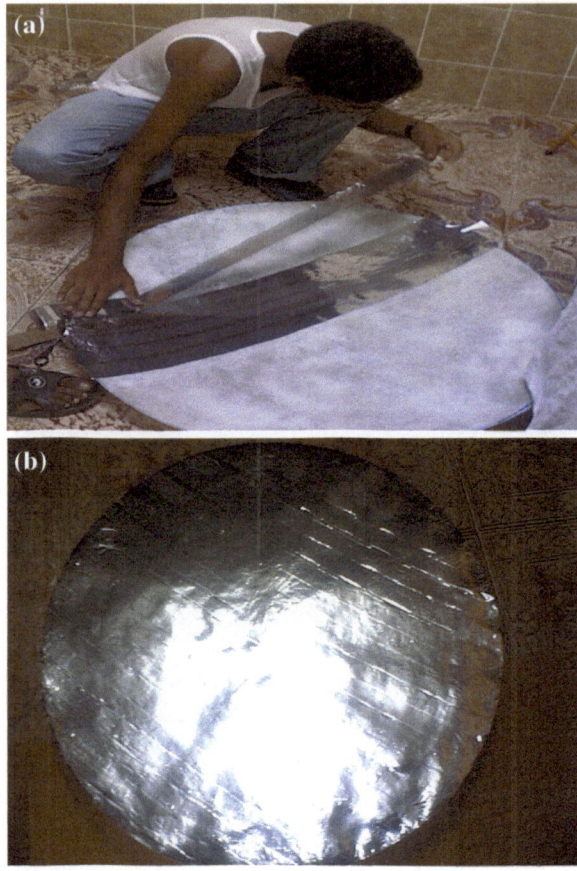

Fig. 4.10 Parabolic dish concentrator after its completion. **a** Gluing the reflective vinyls. **b** Complete parabolic dish

4.7 Chapter Summary

Concentrating solar technologies are in different stages of development. Most of the systems described in this chapter have passed the testing and power production phases in the twentieth century and are being commercialised with each company boasting about their product producing more power at a cheaper price in less space. Yielding the most power per area among all the solar concentrators is the parabolic dish and the latter was selected as our prototype of choice. The steps towards the final parabolic dish concentrator made with fibreglass have been enumerated in the pages of this chapter. To convert the fibreglass parabolic dish to a parabolic concentrator, its surface has to be lined up with a reflective material so as to focus energy optimally. The different types of reflective materials have also been discussed with the pros and cons pointed out for each type. The choice of the solar concentrator along with the building materials (fibreglass and chrome vinyl) were also fully justified in accordance with the ease of production and tight budget.

References

Abbas R, Montes M, Piera M, Martnez-Val J (2012) Solar radiation concentration features in linear fresnel reflector arrays. Energy Convers Manage 54(1):133–144

Chandler D, Ahrens S (2008) Harnessing sunlight on the cheap. In: MIT TechTalk, vol 52. MIT News Office, Cambridge, p 5

Cohen G (2006) Nevada first solar electric generating system. In: IEEE May technical meeting, IEEE, Las Vegas, p 10

Fernandez-Garcia A, Zarza E, Valenzuela L, Perez M (2010) Parabolic-trough solar collectors and their applications. Renew Sustain Energy Rev 14(7):1695–1721

Fishman B (2008) Ausra opens its first concentrating solar power plant in California. In: U.S. Department of Energy: Energy efficiency and renewable energy. U.S. Department of Energy, California, p 4.

FLABEG Holding GmbH (2012) Duraglare solar product information. Datasheet. http://www.flabeg.com/uploads/media/FLABEG_Solar_DuraGlare_02.pdf. Accessed 13 Nov 2012

Harrison J (2001) Investigation of reflective materials for the solar cooker. Florida solar energy center, creating energy independence. http://www.fsec.ucf.edu/en/research/solarthermal/solar_cooker/documents/reflectivematerialsreport.pdf. Accessed 08 Feb 2009

Mancini T, Prairie MR, Kolb G (1997) Advances in solar energy: an annual review of research and development, vol 11, Chap 1. American Solar Energy Society, Inc., Colorado

Price H, Lupfert E, Kearney D, Zarza E, Cohen G, Gee R, Mahoney R (2002) Advances in parabolic trough solar power technology. J Sol Energy Eng 124(109):17

Rabl A (1976) Tower reflector for solar power plant. Sol Energy 18(3):99

ReflecTech, Inc (2009) Reflectech mirror film. ReflecTech product brochure. http://www.reflectechsolar.com/pdfs/ReflecTechBrochuretoEmail22Aug08.pdf. Accessed 08 Mar 2009

Trieb F, Langnib O, Klaib H (1997) Solar electricity generation a comparative view of technologies, costs and environmental impact. Sol Energy 59(1–3):89–99

US Department of Energy (2012a) Dish/engine systems for concentrating solar power. Energy basics. http://www.eere.energy.gov/basics/renewable_energy/images/dish_receivers.gif. Accessed 19 Nov 2012

US Department of Energy (2012b) Linear concentrator systems for concentrating solar power. Energy basics. http://www.eere.energy.gov/basics/renewable_energy/images/parabolic_troughs.gif. Accessed 19 Nov 2012

US Department of Energy (2012c) Linear concentrator systems for concentrating solar power. Energy basics. http://www.eere.energy.gov/basics/renewable_energy/images/linear_frisnel.gif. Accessed 19 Nov 2012

US Department of Energy (2012d) Power tower systems for concentrating solar power. Energy basics. http://www.eere.energy.gov/basics/renewable_energy/images/power_tower.gif. Accessed 19 Nov 2012

Wu S, Xiao L, Cao Y, Li Y (2010) A parabolic dish/amtec solar thermal power system and its performance evaluation. Appl Energy 87(2):452–462

Chapter 5
Results and Discussion

Abstract As a final test, the performance of the tracking solar concentrator in the open field was evaluated. Before jumping to the solar concentrator's efficacy, the validity of the solar tracker's motion was investigated and was in accordance with the actual sun position. Although the parabolic dish was constructed outside a laboratory, the high heat flux at the focussing aperture favoured the ignition/melting of several materials within a few seconds. In an effort to determine the power production of our solar concentrator, a water load, suspended at the focus, was allowed to be heated up rapidly and the power output was computed using various algorithms from the laws of thermodynamics. A constructed model based on the simplified Newton's Law of Cooling could predict the rise in temperature whereby the kinetics of heat transfer are limited by convective cooling. The empirical power output of the 0.6 m^2 tracking solar concentrator was found to be 176 W.

The parabolic dish and the tracker system are two essential components designed to work in perfect synergy. Prior to assembly of these elements together, tests need to be carried out to assess the suitability of each one independently. At a later stage, the operation of the tracker-dish concentrator system is validated.

A considerable span of time was devoted to testing the system due to the high variations in the weather conditions: from torrential rain to clear bright day.

5.1 Tracking Performance

Two methods have been used to ascertain the correct operation of the tracker. In the first method, a straight 20 cm long rod was placed towards the sun perpendicular to the tracker platform (while the tracker is operating) and adjusted manually until the shadow cast by the rod on the platform surface was negligible indicating that the latter was oriented directly towards the sun. The azimuth and altitude angles defined by the rod are compared with the values obtained from the equations in Sect. 2.4. These two sets of values corresponding to different times of two different days are

Z. Jagoo, *Tracking Solar Concentrators*, SpringerBriefs in Energy,
DOI: 10.1007/978-94-007-6104-9_5, © The Author(s) 2013

Table 5.1 The measured angular displacement of an opaque rod pointing directly at the sun as compared to its calculated position

Time	Computed azimuth	Computed altitude	Measured azimuth($\pm2°$)	Measured altitude($\pm2°$)
10 00	101.7°	59.51°	101°	60°
11 00	103.6°	73.26°	103°	73°
12 00	140.2°	86.04 °	140°	87°
13 00	252.9°	78.24°	254°	77°
14 00	258.1°	64.80°	260°	66°
15 00	257.6°	50.80°	257°	52°
16 00	255.5°	37.12°	255°	36°
10 00	101.6°	59.40°	102°	60°
11 00	103.4°	73.15°	105°	73°
12 00	138.6°	86.01°	138°	87°
13 00	253.0°	78.36°	255°	75°
14 00	258.2°	64.91°	260°	66°
15 00	257.7°	50.91°	259°	52°
16 00	255.6°	37.22°	257°	40°

tabulated in Table 5.1. The lack of highly sophisticated equipment rendered measurements imprecise; locally-available protractors can measure up to only a half degree. However, as the rod used had a non-negligible radius, the maximum precision that could be obtained was two degrees. Had a sextant been used, the angular displacement between two far-away objects such as the sun and the horizon could have been measured more accurately (up to 0.2°).

In another experiment, the enhancement of the power output of the panel due to the use of tracking is directly proportional from the measurement results. A high power output will indicate that the tracker is operating with the maximum efficiency while a low power shows that the tracker is not well oriented. A 6 W 12 V (monocrystalline) PV panel of dimensions 260 mm × 157 mm was laid flat on the platform. Starting at 08 00 and after each hour, the open circuit voltage of the PV cell was recorded. After that, cell was removed from the platform and placed on a horizontal plane. The new reading (voltage) in the horizontal position was also recorded. These measurements and observations were carried out throughout the day. The results obtained on Monday 16 March (local sun transit time at 12 32; latitude and longitude at 20°17' S and 57°33' E respectively) were the most reliable due to the exceptionally clear sky conditions that prevailed on that day (Table 5.2).

Visual inspections of the shadow on the platform hinted towards a high degree of alignment with the incoming solar rays. Figure 5.1 illustrates the open *normalized* voltage outputs of the PV cell when it is locked on the tracker platform and when it is on the horizontal ground. At around midday, the enhancement due to the tracker is marginally small and as the sun's altitude angle drops, the open circuit voltage output of the cell is considerably boosted by tracking (36% increase during our test day) which further validates our solar tracker choice in Chap. 3.

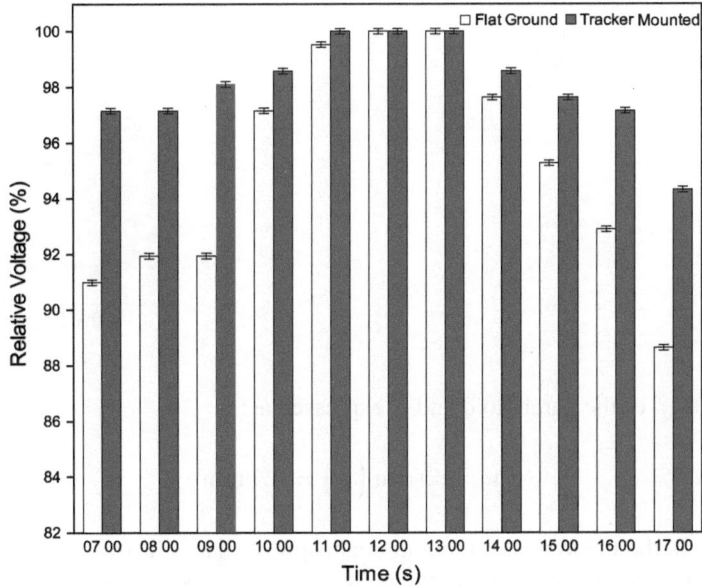

Fig. 5.1 Relative voltage comparison between a tracking PV panel and a fixed PV array

Table 5.2 The voltage measured by the tracker as compared to a fixed system

Time	LCD display		Tracker PV array voltage ($\pm 1V$)	Flat PV array voltage ($\pm 1V$)
	Az	Alt		
07 00	88.01	10.32	20.5	19.2
08 00	82.39	24.33	20.5	19.4
09 00	75.51	38.13	20.7	19.4
10 00	65.59	51.41	20.8	20.5
11 00	48.28	63.27	21.1	21.0
12 00	14.17	70.82	21.1	21.1
13 00	330.1	68.76	21.1	21.1
14 00	303.7	58.95	20.8	20.6
15 00	290.1	46.36	20.6	20.1
16 00	281.7	32.82	20.5	19.6
17 00	275.5	18.90	19.9	18.7

5.2 Study of the Parabolic Dish Focussing Performance

Ideally, the sun's rays are assumed to be perfectly parallel to each other but to an observer on the earth's surface, sun rays emanate from a cone of angular diameter, $\varepsilon = 0.54°$. Therefore, even a perfect parabolic dish cannot give a *reflected* point image of the sun, but a disc of non-zero diameter in the focal plane (Stine and Geyer 2001). The diameter of the image (beam spread) Δd is calculated in terms of the

Fig. 5.2 Diagram showing
the size of the focus with the
theoretical spot at the centre

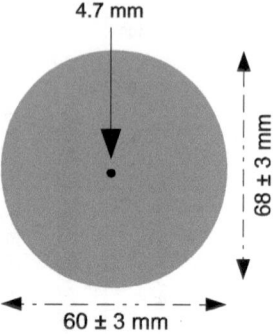

4.7 mm

60 ± 3 mm

68 ± 3 mm

focal length f of the paraboloid and is expressed as:

$$\Delta d = 2f \tan\left(\frac{\varepsilon}{2}\right) = 4.7 \,\text{mm}$$

This value is the theoretical minimum beam spread that will occur. On top of that, any distortion in the shape of the parabolic dish from a perfect one as well as non-specular reflection at the mirror surface contribute to widen the size of the sun's image projected on the focal plane. We attempt to measure the size of the sun's image projected on the focal plane of the parabolic dish by placing a sheet of thermal paper at its focal plane, perpendicular to the dish's axis. The exposure time was chosen so as to selectively darken the thermal paper at the location of hot spots on the focal plane. Figure 5.2 shows a representation of the thermal image obtained. The spot's linear dimensions are about 65 mm, that is 14 times larger than the theoretical size calculated. We attribute this enlarged spot to the presence of deformations in the surface finish due to the fact that the paraboloid was constructed manually with non-specialized tools. Given the limited time frame for the project, we did not attempt to improve the focusing characteristics of the dish. In order to use the largest fraction of the incident solar energy, a receiver that would at least extend 70 mm in diameter needs to be placed at the focal point.

5.3 Temperatures Achieved with Various Materials at the Focus

In an attempt to investigate the working temperatures achievable at the focus of the dish (which is at the centre of the spot measured), several materials were placed there and their temperatures were recorded. In our experiment, the test materials were first blackened to absorb most of the energy and then held at the focus using tongs. The maximum temperature achieved (before any sign of ignition) was determined with the help of an infra-red (IR) pyrometer thermometer (Fluke 80T-IR) having a maximum measurement temperature of 260 °C. Table 5.3 shows the maximum

Table 5.3 Experimental temperature reached for different substances

Material	Process	Temperature (°C)
Plastic A	Melts and ignites	112.1 ± 0.1
Plastic B	Melts	149.8 ± 0.1
Plastic C	Melts	187.2 ± 0.1
60/40 tin-lead Solder	Melts	188.6 ± 0.1
Newspaper	Ignites	218 ± 1
Paper	Ignites	229 ± 1
Wood	Chars and ignites	232 ± 1
Tin	Melts	235 ± 1

recorded temperatures reached by various target materials on a sunny day. Table 5.3 gave us an insight about possible temperature ($\gg 200°$) that the concentrator can maintain on an average sunny day. The temperature at the focus is much more than the measured temperature but the method of placing a solid object at the focus does not represent the actual temperature and thus we have to resort to a more rigorous method to determine the power output of the tracking solar concentrator.

5.4 Prediction of the Available Power of the Solar Concentrator

It is straightforward to gauge the theoretical power output of the solar concentrator. Using an estimate of the solar irradiance, $I_\odot$, the diameter of the dish, d, and the diameter of the receiver, D, the reflectivity of the dish mirror coating, R' and the fraction of energy lost to specular reflection due to unevenness of the mirror surface β, the output power is given by

$$P = I_\odot \times \frac{\pi}{4}(d^2 - D^2) \times R' \times (1 - \beta) \tag{5.1}$$

For $R' = 0.82 \pm 0.05$, $d = 90 \pm 3$ cm and $D = 10 \pm 0.1$ cm, we have

$$P = 0.515(1 - \beta)P_o$$

The value of P can only be an estimate, since no pyranometer was available for measuring the value of $I_\odot$. From Sect. 2.2, we know for a fact that the incident solar irradiance is reduced by about 22% to a maximum of 60% and this is equivalent to a value of $I_\odot$ ranging from 547 to 1066 W/m^2. For the purpose of our calculation, we assume $I_\odot \simeq 750$ W/m^2 and therefore $P \simeq 386$ W if $\beta = 0$.

5.5 Power Output of the Solar Concentrator at the Focal Plane

We next estimated the power available from the concentrator under bright sun conditions using a calorimetric method. Several tests were conducted with water as the load and the effective power output of the system was calculated from temperature measurements.

5.5.1 Measurement Procedure

To determine the power output of the solar concentrator, we placed a metallic can of diameter 120 mm filled with a known volume (150 ml) of water near the focus so that it intercepts the concentrated beam at the focal plane. Three support rods in the form of a tripod were bolted to the parabolic reflector and the can was securely affixed to them as depicted in Fig. 5.3. Since black-bodies are good absorbers of electromagnetic radiation, the tin can was sprayed with a layer of black matt emulsion paint resistant to temperatures above 400 °C. During experiments, the can was covered with a lid having a small perforation at its centre for temperature measurement using the same pyrometer as in the previous section. The role of the lid is to reduce heat losses due to convection currents and water evaporation. After amplification, the temperature output was recorded by a data logger connected to the pyrometer. Because the rate of

Fig. 5.3 Diagram showing a *blackened* can affixed to the concentrating dish using a tripod

Table 5.4 Temperature of 150 ml of water placed at the focus of the solar concentrator measured at 10 s intervals in a calm environment

Time(s)	Temperature (°C)	Time(s)	Temperature (°C)
0	29.2	200	75.3
10	32.6	210	77.2
20	35.4	220	79.4
30	38.0	230	81.0
40	41.0	240	82.6
50	43.4	250	84.0
60	44.9	260	85.1
70	47.1	270	86.9
80	49.3	280	88.0
90	51.7	290	89.4
100	54.3	300	90.1
110	56.1	310	91.1
120	59.2	320	92.9
130	61.0	330	94.0
140	63.7	340	95.1
150	65.6	350	97.0
160	66.7	360	98.2
170	68.5	370	99.1
180	70.9	380	100.0
190	73.4		

temperature change is a rapid phenomenon in this case, the temperature was logged at 10 s intervals.

The experiments were carried out three times to ensure that random errors are eliminated.

5.5.2 Test Day: 15th of March 2009

15th of March 2009 was chosen with malice because of the exceptionally blue sky and lack of bulky clouds which would skew our measurements eventually. In this set-up, the system was placed in a controlled environment suitable for accurate temperature measurements. It was imperative to take measurements around noon because we have deduced from Sect. 5.1 that the tracker is most effective (100 % output power) around that time.

In this experiment, negligible wind was present, so that the heat loss may be considered to be due to natural convection mainly. The temperature profile recorded is displayed in Table 5.4. The associated error with the temperature measurements was 0.1° in all the cases. To enhance the visual perception of the analysis, the data are plotted and the best-fit curve is drawn from the points as illustrated in Fig. 5.4.

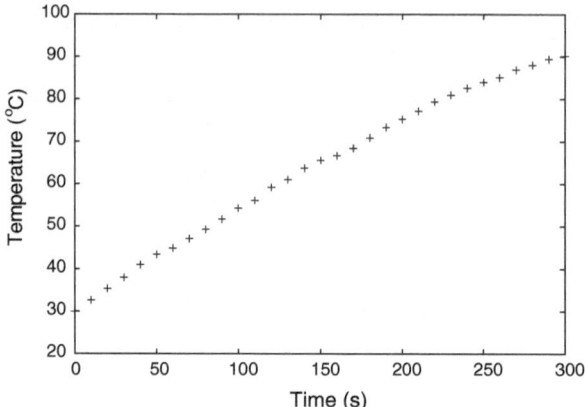

Fig. 5.4 Temperature versus Time for heating water in normal condition

The experiment was started at 12:17 and the water present was allowed to boil.

At the end of the heating time, the water inside the tin can was allowed to cool naturally and its temperature was recorded at regular time periods. Part of the cooling curve is displayed in Fig. 5.5.

The data for this phase was collected over a duration of 41 min at 10 s intervals.

The above experiments underline the fact that the performance of a solar 'heater' might be severely limited by heat losses to the surroundings, particularly under windy conditions.

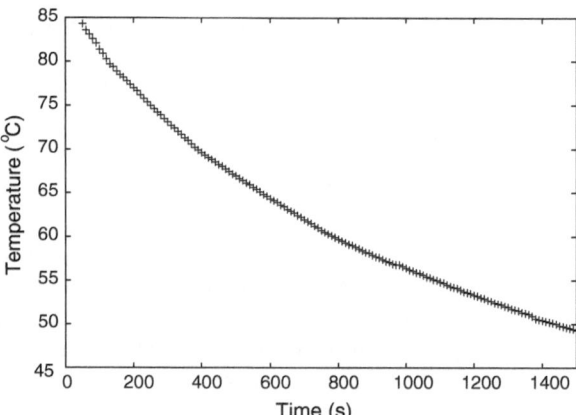

Fig. 5.5 Natural cooling of water (only first 25 min are shown)

5.6 Interpretation of Results

The aim of the experiments was to obtain the output power of our 0.6 m^2 footprint tracking solar concentrator. The previous experiments were performed to achieve that goal. Since the heating rate is of interest to us, the change in temperature per unit time ($\dot{\theta}$) has to be found from Fig. 5.4. This is analogous to the gradient of the curve. In principle, a higher power will produce a higher heating rate and based on this fact, the initial slope needs to be calculated as this gradient corresponds to the heating rate when the heat lost to the surroundings is minimum (Lao and Ramanujan 2004).

The initial slope corresponding to $\dot{\theta}$ was calculated from the manipulation in Fig. 5.6 and was found to be 0.270 ± .005 °C/s. Since all the experiments were repeated at least twice, the statistical mean was performed to reduce the overall (random) error associated with the experiments. The standard deviation of this distribution was calculated using

$$\sigma = \sqrt{\frac{1}{N} \sum_{i=1}^{N} (x_i - \mu)^2}, \quad \text{where} \quad \mu = \frac{1}{N} \sum_{i=1}^{N} x_i \qquad (5.2)$$

and the standard error in the rate of heating is obtained from

$$\delta \dot{Q} = \frac{\sigma}{\sqrt{N}} \qquad (5.3)$$

where N is the number of times each experiment was performed.

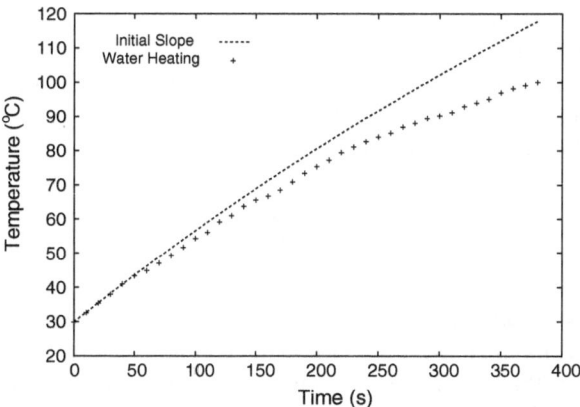

Fig. 5.6 Temperature versus Time with initial slope for water heating

From thermodynamics, the heating rate can be calculated from

$$\dot{Q} = mc_p \dot{\theta} \qquad (5.4)$$

where

$\dot{\theta}$ is the rate of temperature rise (°C/s),

$\dot{Q}$ is the input power (J/s or W),

m is the mass of water (kg),

c_p is the heat capacity of water at constant pressure (J/kg°C).

The maximum heating rate, $\dot{Q}$ and subsequently the output power, P was found to be 169.3 ± 3.1 W.

A slightly different approach which will theoretically yield the same result is to use Newton's Law of Cooling/Heating and model the experimental curves. This method is by far more accurate when a fluid is being heated while energy is escaping the fluid to the surroundings since it takes into account the minute losses in real-time. This is not the case with our previous method although we assume that the heat loss to the surroundings is minimum initially but this is a far-fetched approximation between subsequent 10 s measurements.

The rate of temperature rise depends on the difference between the heat input to the load minus the heat losses modelled on Newton's Law of Cooling.

$$\dot{\theta} = \frac{\dot{Q}}{mc_p} - \left(\frac{\alpha}{mc_p}\right) \times (\theta - \theta_o) \qquad (5.5)$$

where

θ is the temperature of the water (°C),

θ_o is the room temperature (°C), and

α is the Newtonian constant (J/°Cs).

The Newtonian constant is dependent upon the geometry and the material of the recipient in which the liquid is heated or cooled.

Upon solving this equation, the temperature turns out to be

$$\theta = \theta_o + \frac{\dot{Q}}{\alpha} \left(1 - e^{\frac{-\alpha t}{mc_p}}\right) \qquad (5.6)$$

for temperatures (θ) which are less than the boiling point of water.

The equation has too many variables to be solved immediately and has to be solved indirectly. This method involves extracting the required information from the cooling data. When a fluid is being cooled, Eq. 5.5 is modified as follows (O'Sullivan 1990)

$$\dot{\theta} = -\left(\frac{\alpha}{mc_p}\right) \times (\theta - \theta_o). \qquad (5.7)$$

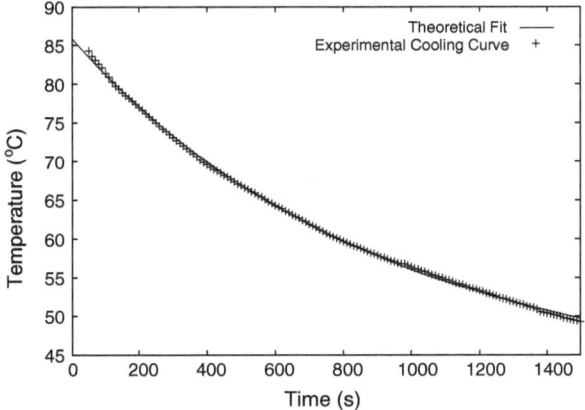

Fig. 5.7 Cooling curve for water

Being a separable first-order differential equation, Eq. 5.7 can be solved rather easily and produces the following solution

$$\theta = \theta_o + (\theta_I - \theta_o)\, e^{\frac{-\alpha t}{mc_p}} \qquad (5.8)$$

where θ_I is the temperature where cooling starts.

The constant α can be determined by fitting Eq. 5.8 to the experimental cooling data.

$\theta_I = 84\,°C$ and $\theta_o = 29.9\,°C$ are experimental constants and as we can deduce, the theoretical fit in Fig. 5.7 is quite accurate with a residual error of less than $2\,°C$ for the first 25 minutes of cooling, only diverging at the beginning and end points. Thus, the theoretical estimate and the actual measured values were in good correspondence.

After α has been obtained, it is in turn plugged into Eq. 5.5 and an attempt to fit the heating data of Fig. 5.4 was made by using the values obtained previously. It is remarkable that the value of α determined from the cooling curve gives very close matching with the experimental curve. The quality of the fit can be witnessed from Fig. 5.8.

The value of the coefficient α is found to be $1.46 \pm 0.04\,J/°Cs$ when $\theta_o = 29.9\,°C$, $m = 0.15\,kg$ and $c = 4180\,J/kg\text{-}°C$ constants are employed.

The essence of fitting the theoretical cooling/heating data to the experimental curves was to find the numerical value of the output power. From our experiments, the concluded value of $\dot{Q}$ was found to be $176.0 \pm 1.5\,J/s$ and this number corresponds to the power of the solar concentrator (P) at full-capacity.

It is important to point that the Newton's model can predict quite accurately the rate of temperature rise in the more or less ideal conditions. However, at high temperatures, the model cannot predict accurately the change of temperature with time. This is explained by the fact that at higher temperatures, the rate of evaporation

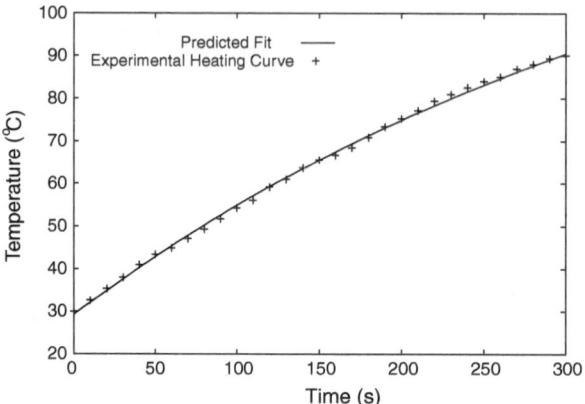

Fig. 5.8 Prediction of the heating curve during normal conditions

becomes significant and is maximum at the boiling point of water. With evaporation, molecules with high internal energy escapes the liquid and the overall temperature of the liquid decreases. To work out this problem, another fluid having a higher boiling point like oil can replace the water. When the environment of the metallic can changes (absence of lagging, presence of wind, etc.) the law is still approximately valid, provided the value of α is modified.

Two different methods to calculate the output power of a solar concentrator were utilized and using Newton's Law of Cooling as a model produced a more accurate result. Newton's Law of Cooling took into account all the minute losses while heat is being injected and the eventual power of the solar concentrator was found experimentally to be in the vicinity of 176.0 W.

The concentrator has a working efficiency of 45 % when compared to the theoretical estimate of the solar power of 386 W from Sect. 5.4. The lower than expected efficiency can be explained by specular reflection at the mirror surface, diffused reflection on load, deviation of the dish geometry from an ideal paraboloid of revolution, and rough parabolic surface due to construction flaws.

5.7 Chapter Summary and Prospects

As a final test, the performance of the tracking solar concentrator in the field was evaluated. The various tests comprise of validating the tracking motion of the solar tracker and testing whether some materials like wood, paper etc. could be ignited when a sufficiently high heat flux is incident on them. Ultimately, in an effort to determine the experimental power production of our solar concentrator, a water load, suspended at the focus, was allowed to be heated up rapidly and the power output was computed using various algorithms from the laws of thermodynamics. The kinetics of the heat transfer are limited by convective cooling. This type of cooling has been shown to conform to Newton's Law of Cooling.

The heating of water can be realised on a continuous flow basis by maintaining a circulation of water through a receiver placed at the focus. Another interesting option would be to produce electricity by means of PV cells capable of operating under high irradiance or through compact Stirling engine.

References

Lao L, Ramanujan R (2004) Magnetic and hydrogel composite materials for hyperthermia applications. J Mater Science: Mater in Medicine 15(10):1061–1064

O'Sullivan C (1990) Newtons law of cooling—a critical assessment. Amer J Phys 58(10):956–960

Stine W, Geyer M (2001) Power from the Sun, 2nd edn. California State Polytechnic University, California

Chapter 6
Conclusion

Abstract An overview of what has been achieved in this book is presented in this chapter. Starting from the position of the sun to the prototyping of the solar tracker and culminating at the construction of the solar concentrator is briefly explained. Future steps that can enhance the robustness of the solar tracker, the perfection of the solar paraboloid and the improvement of the reflective material are also exhibited. Ultimately, details about a braking mechanism as means to reduce the power consumption of the 30 W solar tracker is lucubrated.

We have presented a synthesis of our works on a solar energy concentration device with the ability to track the sun for optimum performance. This multi-disciplinary study has taken us on an interesting journey across various fields like astronomy, mechanical engineering, electrical and electronics engineering, optics and thermo-dynamics.

Our work was staged according to three distinct phases.

We have studied the sun's position in the celestial sphere and the main factors governing it. This has allowed us to formulate a coherent mathematical procedure to evaluate the altitude and azimuth angle of the sun with the required degree of precision. This was followed by efforts to code this procedure in the limited memory of a PIC® microcontroller. Moreover various support circuits, peripherals, interfacing input/output devices and electromechanical devices were annexed to the microcontroller which controlled and orchestrated their operations so that the end result was a working solar tracker.

The construction of a solar concentrator dish was the second task that we undertook. It's design involved studying the dish geometry, the mirror surface film and the mechanical attachment of the device to the tracker platform. Fibreglass was chosen as the dish material as it fulfilled the requirements of strength, sturdiness, lightness and low cost that we had set. Although, it was far from ideal, a vinyl film mirror was applied to the dish as it was available locally in a timely manner. It did a good job for evaluation purposes, but it was clear that future designs would need to offer better characteristics.

Z. Jagoo, *Tracking Solar Concentrators*, SpringerBriefs in Energy,
DOI: 10.1007/978-94-007-6104-9_6, © The Author(s) 2013

   The last phase involved the study of the performance of the tracker-concentrator. The concentrator focussed solar energy into a spot of diameter of less than 65 mm. Tracking performance was assessed using two methods and proved extremely effective at boosting the performance of a PV cell mounted on the tracker. The power delivered by the tracker-concentrator to a water load was evaluated as 176 ± 1.6 W. The effect of convective heat losses was found to be significant and we recommend that it should be minimised, by lagging or enclosing the load inside a transparent glass (evacuated) enclosure.

   This work sets the foundation for a sophisticated solar concentrator that is versatile enough to receive a host of receivers from simple water loads to more sophisticated off-the-shelf (or experimental) Stirling engines or high irradiance PV cells.

6.1 Recommendations for Further Work

The system in itself is very challenging but there is still room for improvement.

   The realised units are power demanding, actually, the system while operating consumes on average 30 W and the majority of this power is devoted to the motors. The design is to be reviewed so that the system runs on less power. To resolve this problem, one can use a braking mechanism which will lock the platform holding the dish intact (Fig. 6.1). The brake will activate as soon as the stepper motor has oriented the parabolic dish and when the latter is buckled, the stepper motor is switched off. Unlocking will take place just before the microcontroller sends the appropriate pulses to the motors for proper positioning towards the sun.

   The parabolic dish constructed could not focus the incident sun beam to the theoretical 4.7 mm spot because it was made using traditional methods in a short

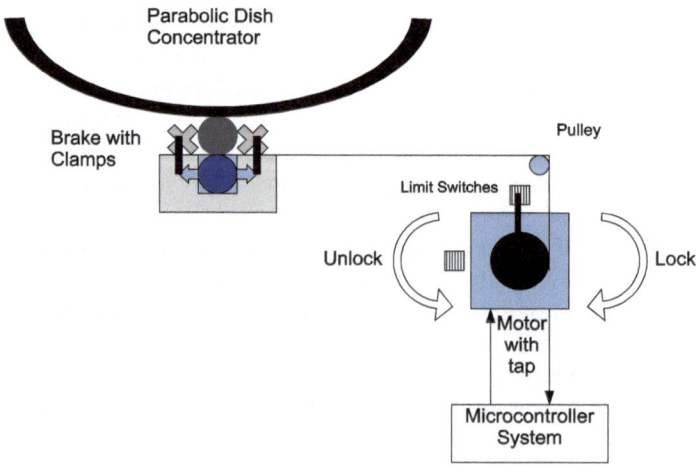

Fig. 6.1 Braking mechanism for the tracker system

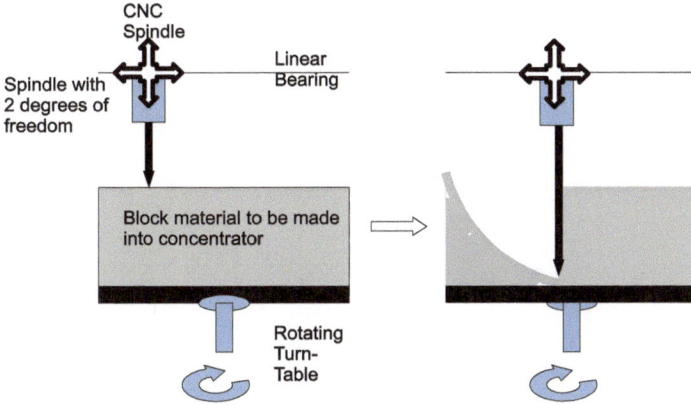

Fig. 6.2 CNC layout for making a parabolic concentrator

time period. However if a CNC machine operating on a turn-table was available like in Fig. 6.2, then resulting concentrator would have been to stricter tolerances and hence reduce specular reflection. The principle of operation of this machine is elementary: a block to be milled is placed on a rotating table and a high frequency spindle drills the material at programmed depths. The turn-table homogenises the material at all points from the edge of the block.

The reflective material glued to the parabola had a reflectivity of only 82 % and upon pasting, there were some inconsistencies due to the overlapping of the vinyl resulting in the creation of bubbles. This diminished the concentrating power and instead, large rolls of commercial reflective films like Reflectech which costs $3/ft^2 could be imported from the USA and used in future concentrating dishes. These have the advantages that they have a high reflectivity (>94 %) and can be cut into triangular stripes that almost nullifies the overlapping of films, hence boosting the performance of the concentrator.

Appendix A
Matlab\Octave Script for Computing the Sun's Position

```
%Input the Date for which the Sun Azimuth Position Data is to be
generated.
% Date Format : dd-mm-yyyy
opt_prompt = input('Should the data be generated for today [y/n] : ','s');
if isempty(opt_prompt) == 1
      return
end
if opt_prompt == 'n',
      data_date = input('Enter the date in MM-DD-YYYY : ','s');
      if isempty(data_date) == 1
          return
      end
      % Serial Date Number from 1-Jan-2000
      data_date_sdn = datenum(data_date);
      % Date components
      date_mat = datevec(data_date_sdn);
elseif opt_prompt == 'y',
      date_mat = clock;
      data_date = datestr(date_mat,1);
else
      disp('You were allowed only to press y or n. Script terminated.')
      return
end
% 'datentime' holds the value of starting time.
Year = date_mat(1);
Month = date_mat(2);
Day = date_mat(3);
% Input the Starting Time, Number of Intervals and Duration of Interval
for which the Sun Azimuth Position Data is to be generated.
data_time = input ('Enter the Starting Time in HH:MM:SS [24 Hours] format
: ','s');
if  isempty(data_time) == 1
      return
end
data_time_sdn = datenum(data_time);
time_mat = datevec(data_time_sdn);

Hour = time_mat(4);
Minute = time_mat(5);
```

Z. Jagoo, *Tracking Solar Concentrators*, SpringerBriefs in Energy,
DOI: 10.1007/978-94-007-6104-9, © The Author(s) 2013

```
Seconds = time_mat(6);
UT = (Hour + (Minute/60)) - 4;
d = 367*Year - fix(7*((Year + fix((Month+9)/12))/4)) +
fix((275*Month)/9) + Day - 730530 + UT/24;
oblecl = 23.4393 - 3.563E-7 * d;
w = 282.9404 + 4.70935E-5 * d;
MA = 356.0470 + 0.9856002585 * d;
e = 0.016709 - 1.151E-9 * d;
E = MA + e*(180/pi) * sind(MA) * ( 1.0 + e*cosd(MA));
A = cosd(E) - e;
B = sqrt(1 - e*e) * sind(E);

v = atan2d( B, A );
slon = v + w;

RA = atan2d((sind(slon) * cosd(oblecl)), cosd(slon));
RA = rev(RA);
Dec = asind(sind(oblecl) * sind(slon));

long = +57.55; lat = -20.28333333;

LST = 98.9818 + 0.985647352 * d + long + UT*15;
LST = rev(LST);
H = LST - RA;
H = rev(H);
Altitude = asind(sind(lat) * sind(Dec) + cosd(lat) * cosd(Dec) * cosd(H))
Azimuth = atan2d(sind(H),(cosd(H) * sind(lat) -
tand(Dec) * cosd(lat))) + 180

LST0 = 98.9818 + 0.985647352 * d + long - 4*15;
LST0 = rev(LST0);

MT = RA - LST0;
if MT < 0,
     MT = MT + 360;
end

h0 = -1*50/60;
H0 = acosd((sind(L0) - sind(lat) * sind(Dec))/(cosd(lat) * cosd(Dec)));
H0 = rev(H0);

SunriseHr = 0;
SunriseHr = 0;
Sun_Rise = MT - H0;
Sunrise = Sun_Rise/15;

SunriseVar = Sunrise
     while SunriseVar > 1,
     SunriseVar = SunriseVar - 1;
          SunriseHr = SunriseHr + 1;
     end
SunriseHr;
     SunriseMn = SunriseVar * 60;
```

Appendix B
Solar Tracker Circuit Realisation

B.1 Gerber Format

The schematic as depicted in Fig. 3.18 is drawn in a PCB design software, in our case, FreePCB 1.358 and the top, bottom and drill layers of PCB map is converted using the in-built function into the common Gerber file format (*$g.br$). The reason why this additional conversion is absolutely necessary is because the Gerber format is the principle standard format used in manufacturing industry for PCB designs, regardless of which propriety format the PCB is saved by the designing software. The Gerber layers for the circuit is shown in Fig. B.1.

B.2 Milling of Gerber Layers

Although the Gerber format is regarded as a standard, it is yet not ready to be processed till another conversion is effected. IsoCAM, a software accompanying the Bungard CNC machine, takes the Gerber files and creates the milling data by isolating the copper traces, thereby reducing the probability of occurrence of shorts. After the Gerber files have been loaded into IsoCAM 2.0, the next step is to change the drill tools according to the components used and the drilling tools available with the Bungard CNC machine.

It is wise to do the DRC (Design Rule Check) available in IsoCAM on both the top and bottom layers. A step, not to be missed, is the mirroring of the top, bottom and drill layer and this is done by faking a vertical fixing hole with no apparent diameter and distance from the outline of the circuit. Since we are dealing with two-sided sided PCBs, it is imperative to synchronise the top and bottom layer and to do so, we add a horizontal fixing hole with a diameter of 2.5 mm (thickness of fixing pins that are supplied with the machine) and a distance of at least 5.0 mm to the circuit so that depth limiters could be used. We should then see two horizontal fixing holes appear on the extremities of our design and we mirror the bottom layer *only*. The final step

Z. Jagoo, *Tracking Solar Concentrators*, SpringerBriefs in Energy,
DOI: 10.1007/978-94-007-6104-9, © The Author(s) 2013

Fig. B.1 Gerber layers for the
Solar Tracker. **a** Top Gerber
layer. **b** Bottom Gerber layer.
c Drill Gerber file. **d** All files
superimposed on each other

(a)

(b)

(c)

(d)

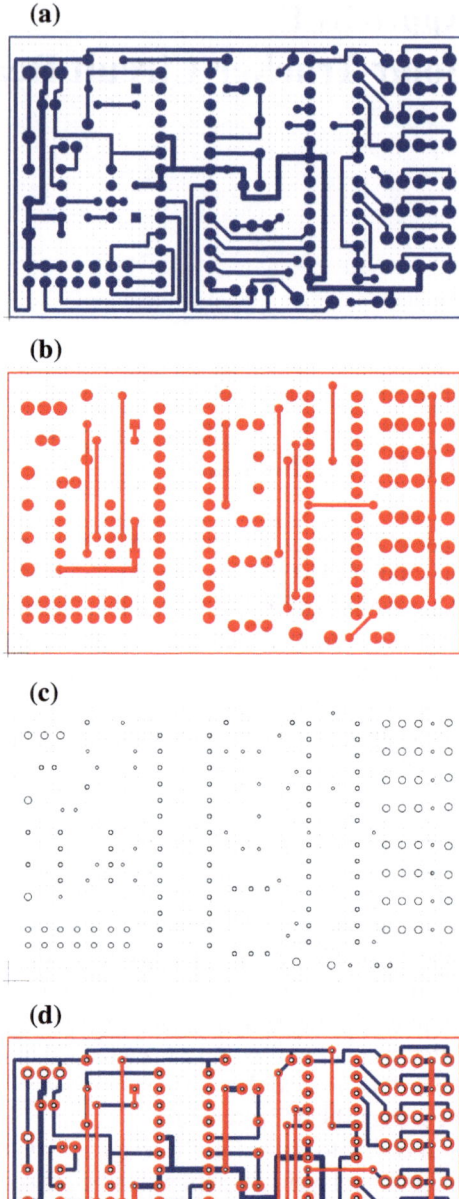

Fig. B.2 Milled diagram of tracker circuit

Table B.1 Table showing the drill tools used for different components

#	Component	Diameter of drill tool used (\mm)
1	Through-Hole	0.6
2	Crystal oscillators	0.6
3	IC holders	0.8
4	Capacitors	0.8
5	LCD	1.0
6	Resistors	1.0
7	HEXFETs	1.2
8	7805 Regulator	1.2
9	Power rails	1.2

consists of creating the milling data of the top and bottom layers (shown in Fig. B.2). We should take note that it is important to set the options with reference to the manual of the CNC machine—for the milling parameters:

1. we set the tolerance to be 0.0099 mm and
2. the output resolution, 0.025 mm (HPGL).

The milling tool depends usually on the accuracy of the PCB designed and in our case, a milling tool of 0.2 mm (Table B.1) was used since the PCB was optimised for its size (70 × 40 mm). Milling data is saved with a 'plt' extension and drilling data (drill data and fixing holes), with a 'ncd' extension. The extension must at all cost be included when saving for IsoCAM does not automatically append an extension to the file name. These steps marks the end of the conversion processes and a blank circuit board is ready to be etched (Table B.1).

B.3 Circuit Realisation Using RoutePro and DrillPro

A Computer Numeric Control (CNC) machine (Bungard CCD Manual Tool Change (MTC)) was used to etch the circuit on a piece of copper board. First, a proper offset position is defined so that it will start at the correct location on the circuit board. Fixing holes of the same diameter as the fixing pins are drilled using DrillPro so that the board can be secured firmly to the MDF base board. The stroke for these holes should be deep enough so that the pins are upright and tight. Then, we make sure the board and base thickness are correct so that routing and drilling processes have pin-point precision.[1] Then, we move to RoutePro and load the bottom HPGL file. It is of extreme importance to use the same offset settings as before. Since we are dealing with a two-sided PCB, we have to route both the top and bottom layers and after the bottom layer has been routed, we turn the circuit board over to the other side by flipping it around the x-axis and place it back onto the base board. As for the bottom layer, the top layer was etched by loading the top HPGL file in RoutePro. To finish the process, we return back to DrillPro and drill the holes for each and every component on the top layer with the same initial settings (offset, base board thickness and board size dimensions) by loading the appropriate drill file.

B.4 Advantages of Using CNC Machines Over Traditional Methods

- Speed—The solar tracker circuit was manufactured in one hour, routing and drilling inclusive compared to 10 days if the circuit had been manufactured using a photo-etching method.
- Ease—It would have been next to impossible to make a two-sided PCB with UV radiation and ferric chloride solution as it is very difficult to synchronise the top and bottom layers. However, with desktop manufacturing, a pair of fixing pins does the job.
- Reproducibility—It is very easy to reproduce the circuit once the HPGL files have been generated and any number of identical boards can be cloned.
- Ease of modifying circuit—Changes can be brought by modifying the layout in the PCB design software and regenerate the necessary HPGL files using IsoCAM.

---

[1] Note: We should stand clear of the milling table after drilling begins as moving parts may cause personal injury or damage to the machine if they are touched.

Appendix C
PIC Microcontroller ASM Code

```
; The code written shows the interfacing between the microcontroller
; and Hitachi HD44780 controller compatible LCD only.

; Wiring:
;    LCD 0-7 lines are wired to PIC RD0-RD7
;    LCD E, RS & RW lines are wired to PIC RE1, RE0 & RE2
;    Motor A HEXFET 1 & 2 lines are wired to PIC RC0 & RC1
;    Motor B HEXFET 1 & 2 lines are wired to PIC RC2 & RC3
;    DS1307 SDA line is wired to PIC RB2
;    DS1307 SCL line is wired to PIC RB1

   processor 18F2682          ; Define processor
   include <p18F2682.inc>   ; Header file for PIC18F2682
   __CONFIG _CP_OFF & _WDT_OFF & _BODEN_OFF & _PWRTE_ON &
   _XT_OSC & _WDT_OFF & _LVP_OFF & _CPD_OFF
; __CONFIG directive is used to embed configuration data.
;===========================================================;
; PIC-to-LCD pin wiring and LCD line
;===========================================================;
#define    E_line  1         ;|
#define    RS_line 0         ;| -- from wiring diagram
#define    RW_line 2         ;|
; LCD line addresses (from LCD data sheet)
#define    LCD_1 0x80        ; First LCD line constant
#define    LCD_2 0xc0        ; Second LCD line constant
;===========================================================;
;   General Purpose Variables and Local Equates
;===========================================================;
; Reserve 20 bytes for string buffer
   cblock  0x20
     strData
   endc
LCD_Counter equ 0x37                  ; LCD counter
pic_ad equ 0x39                       ; Storage for start of text
index equ 0x3A                        ; Index into text table
TMP_0 equ 0x3B                        ; Temporary register 0
YEAR equ 0x41                         ; Year register
```

```
MONTH equ 0x42                          ; Month register
DOW equ 0x43                            ; Day of week register
DAY equ 0x44                            ; Date register
HOURS equ 0x45                          ; Hour register
MINUTES equ 0x46                        ; Minutes register
SECONDS equ 0x47                        ; Seconds register
DAY_CHECK equ 0x4D                      ; Day change check register
SUNRISE_HOURS equ 0x53                  ; Sunrise hour register
SUNRISE_MINUTES equ 0x54               ; Sunrise minute register
Alt equ 0x55                            ; 4 bytes long altitude
Az equ 0x59                             ; 4 bytes long azimuth


;===============================================================;
;     P R O G R A M
;===============================================================;
org   0x00                             ; start at address
; We set PORT A and B for output
; By default port A lines are analog. To configure them
; as digital we must set bits 1 and 2 of the ADCON1 register.
  SET_BANK1
  movlw 0x07
  movwf CMCON
  movlw 0x07
  movwf ADCON1
  movlw b'00000000'                    ; All lines to output
  movwf TRISA                          ; in port A
  movwf TRISB                          ; and port B
; We wait and initialize HD44780 LCD
  call  delay_5ms                      ; Allow LCD time to initialize
  call  initLCD                        ; Then do forced initialization
; We store base address of text buffer in PIC RAM
  movlw 0x20                           ; Start address of text
  movwf pic_ad                         ; buffer to local variable
DAY_CHANGE
  call  LCD_LINE1                      ; Routine to check whether the
  movf  DAY_CHECK, W                   ; day has changed. If not, loop
  xorwf DAY, W                         ; till then. Tracking starts on
  btfsc status, Z                      ; a fresh day to prevent damage
     goto DAY_CHANGE                   ; of motors
; Complex routine (not listed here) to do all the calculations
; of the position of the sun and then displays it on the LCD.
; it saves the altitude, azimuth and the sunrise in memory.
; The motors are then positioned. This is an open-loop.
;***************************************************************
; INITIALIZE LCD PROCEDURE
;***************************************************************
initLCD
; Initialization for Densitron LCD module as follows:
; 8-bit interface, 2 display lines of 20 characters each
; cursor on, cursor shift right, left-to-right increment
;***********************
; COMMAND MODE
;***********************
```

```
    bcf    PORTA,E_line              ; E line low
    bcf    PORTA,RS_line             ; RS line low for command
    bcf    PORTA,RW_line             ; Write mode
    call   delay_125mcs              ; Delay 125 microseconds
;***********************
; FUNCTION SET
;***********************
    movlw  0x3B                      ; 0 0 1 1 1 0 0 0 (FUNCTION SET)
                                     ;         | | | |_ font select:
                                     ;         | | | 1 = 5x10 in 1/(8/11)
                                     ;         | | | 0 = 1/16 dc
                                     ;         | | |___ Duty cycle select
                                     ;         | | 0 = 1/8 or 1/11
                                     ;         | | 1 = 1/16
                                     ;         | |___ Interface width
                                     ;         |   0 = 4 bits
                                     ;         |   1 = 8 bits
                                     ;         |___ FUNCTION SET COMMAND
    movwf  PORTB                     ; Send data to LCD lines
    call   pulseE                    ; PulseE and delay
;***********************
; DISPLAY AND CURSOR ON
;***********************
    movlw  0x0C                      ; 0 0 0 0 1 1 0 0 (DISPLAY)
                                     ;         | | | |___ Blink
                                     ;         | | | 1 = on, 0 = off
                                     ;         | | |___ Cursor on/off
                                     ;         | | 1 = on, 0 = off
                                     ;         | |____ Display on/off
                                     ;         | 1 = on, 0 = off
                                     ;         |_____ COMMAND BIT
    movwf  PORTB                     ; Send data to LCD lines
    call   pulseE                    ; PulseE and delay
;***********************
;  ENTRY MODE SET
;***********************
    movlw  0x06                      ; 0 0 0 0 0 1 1 0 (ENTRY MODE)
                                     ;           | | |___ display
                                     ;           | | 1 = shift
                                     ;           | | 0 = no shift
                                     ;           | |___ cursor
                                     ;           | 1 = left-to-right
                                     ;           | 0 = right-to-left
                                     ;           |___ COMMAND BIT
    movwf  PORTB                     ; Send data to LCD lines
    call   pulseE                    ; PulseE and delay
;***********************
; CURSOR/DISPLAY SHIFT
;***********************
    movlw  0x14                      ; 0 0 0 1 0 1 0 0
                                     ;         | | | |_|___ don't care
                                     ;         | |_|__ cursor/display
                                     ;         |  00 = cursor shift left
```

```
                                         ;          | 01 = cursor shift right
                                         ;          | 10 = cursor and display
                                         ;          | shifted left
                                         ;          | 11 = cursor and display
                                         ;          |     shifted right
                                         ;          |__ COMMAND BIT
      movwf    PORTB                     ; Send data to LCD lines
      call     pulseE                    ; PulseE and delay
      call     delay_5ms                 ; More delay
;==============================
; LCD display procedure
;==============================
; Sends 16 characters from PIC buffer with address stored in
; variable pic_ad to LCD line previously selected
display20:
; Set up for data
   bcf    PORTA,E_line                   ; E line low
   bsf    PORTA,RS_line                  ; RS line low for control
   call   delay_125mcs                   ; Delay
; Set up counter for 16 characters
   movlw D'20'                           ; Counter = 16
   movwf LCD_Counter
; Get display address from local variable pic_ad
   movf   pic_ad,w                       ; Display RAM address to W
   movwf  FSR                            ; W to FSR
getchar:
   movf   INDF,W
; Get character from display RAM location pointed to by file
; select register (FSR)
    movwf    PORTB                       ; Output to port B
    call     pulseE                      ;send data to display
; Test for 20 characters displayed
    decfsz  LCD_Counter,f                ; Decrement counter
    goto    nextchar                     ; Skipped if done
    return
nextchar:
    incf    FSR,F                        ; Bump pointer
    goto    getchar                      ; Loop
;=========================
; Set address to LCD line 1
;=========================
; ON ENTRY:
; Address of LCD line 1 in constant LCD_1
line1:
   bcf    PORTA,E_line                   ; E line low
   bcf    PORTA,RS_line                  ; RS line  low = control
   call   delay_125mcs                   ; delay 125 microseconds
   movlw  LCD_1                          ; Set to display line
   movwf  PORTB
   call   pulseE                         ; Pulse and delay
   bsf    PORTA,RS_line                  ; Set RS line for data
   call   delay_125mcs                   ; Delay
   return
```

```
;================================
; first text string procedure
;================================
storeDate:
; Procedure to store in PIC RAM buffer the message contained
; in the code area labelled msg1.
; ON ENTRY:
; pic_ad holds address of text buffer in PIC RAM and w
; hold offset into storage area. msg1 is routine that returns
; the string characters and a zero terminator index is local
; variable that hold offset into text table. This variable is
; also used for temporary storage of offset into buffer.
; ON EXIT:
; Text message stored in buffer
; Store offset into text buffer (passed in the w register) in
; temporary variable
movwf    index                     ; Store w in index
; Store base address of text buffer in FSR
movf     pic_ad, W                 ; Display RAM address to W
addwf    index,W                   ; Add offset to address
movwf    FSR                       ; W to FSR
; Initialize index for text string access
movlw    0x00                      ; Start at 0
movwf    index                     ; Store index in variable w
; still index = 0
get_msg_char:
call     msg1                      ; Get character from table
andlw    0x0FF                     ; Test for zero terminator
btfsc    STATUS,Z                  ; Test zero flag
  goto     endstr1                 ; End of string
; ASSERT: valid string character in w store character in text
; buffer (by FSR)
movwf    INDF                      ; Store in buffer by fsr
incf        FSR,f                  ; Increment buffer pointer
; Restore table character counter from variable
movf     index,W                  ; Get value into w
addlw    0x01                      ; Bump to next character
movwf    index                    ; Store index in variable
goto     get_msg_char             ; Continue
endstr1:
return
; Routine for returning message stored in program area
msg1:
addwf    PCL,F                     ; Access table
addlw    DATE
movwf    TMP_0
rlf      known_zero, W
addlw    high(DATE)
movwf    PCLATH
movf     TMP_0,W
movwf    PCL
DATE     retlw    ' '
retlw    ' '
```

```
retlw    ':'
; Each character is  represented by retlw till the 20th character.
retlw    0x00
;This table displays date and time in the format HH:MM:SS DD/MM/YYYY.
;=====================
; first LCD line
;=====================
; Store 20 blanks in PIC RAM, starting at address stored
; in variable pic_ad
LCD_LINE1:
call    delay_125mcs               ; Wait for termination
; Call procedure to store ASCII characters in text buffer
movlw   d'1'                       ; Offset into buffer
call    storeDate                  ; Display the date table
movf    HOURS, W                   ; Convert HOURS variable
call    BCD2ASCII                  ; into two ASCII values
movf    ASCII_TENTH, W             ; for LCD.
movwf   0x21                       ; Position 1 on LCD
movf    ASCII_UNIT, W              ; This goes on till position 20
; Set DDRAM address to start of first line
call    line1
; Call procedure to display 16 characters in LCD
call    display20
return
;=====================
; Pulse E line
;=====================
pulseE
bsf     PORTA, E_line              ; Set E line
bcf     PORTA, E_line              ; Clear E line
call    delay_125mcs               ; Delay 125 microseconds
return
```

Appendix D
Mechanical Systems

D.1 Pulley Systems

Generally stepper motors move by a fixed angular displacement per step and we will consider only motors having specification 1.8°/step.

For one step of this motor:

$$\text{Length of belt passing over driver} = \frac{1.8°}{360°} \times \pi d_1 \qquad (\text{D.1})$$

$$\text{Length of belt passing over follower} = \frac{\theta}{360°} \times \pi d_2 \qquad (\text{D.2})$$

Since belt is taut at all times (Fig. D.1),

$$\frac{\theta}{360°} \times d_2 = \frac{1.8°}{360°} \times d_1 \qquad (\text{D.3})$$

$$\therefore \frac{d_1}{d_2} = \frac{\theta}{1.8°} \qquad (\text{D.4})$$

$$\text{if } d_2 = 3d_1 \qquad (\text{D.5})$$

$$\therefore \theta = 0.6° \qquad (\text{D.6})$$

$$\tau_{driver} = T_1 r_1 - T_2 r_1 \qquad (\text{D.7})$$

$$\tau_{follower} = T_1 r_2 - T_2 r_2 \qquad (\text{D.8})$$

$$\frac{\tau_{follower}}{\tau_{driver}} = \frac{(T_1 - T_2)r_2}{(T_1 - T_2)r_1} \qquad (\text{D.9})$$

$$\therefore \tau_{follower} = \frac{d_2}{d_1}\tau_{driver} \qquad (\text{D.10})$$

$$= 3\tau_{driver} \qquad (\text{D.11})$$

Z. Jagoo, *Tracking Solar Concentrators*, SpringerBriefs in Energy,
DOI: 10.1007/978-94-007-6104-9, © The Author(s) 2013

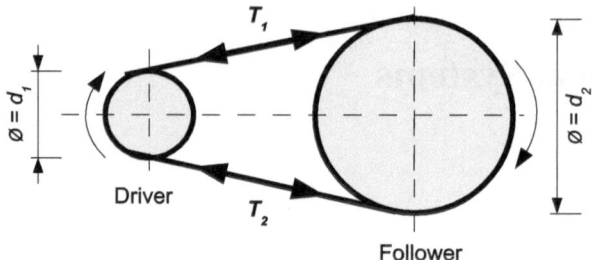

Fig. D.1 Diagram showing a driver and a follower

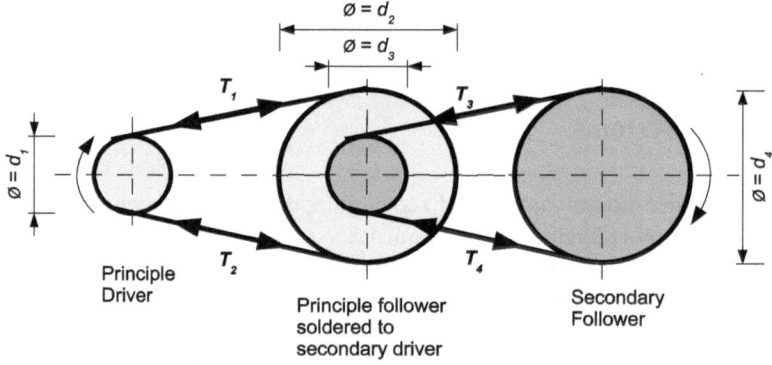

Fig. D.2 Diagram showing two pulley systems soldered together

Taking the principle driver and the secondary driver as independent systems, For one step of principle driver as shown in Fig. D.2:

$$\frac{d_1}{d_2} = \frac{\theta}{1.8°} \tag{D.12}$$

$$\therefore \theta = 0.6° \text{ if } d_2 = 3d_1 \tag{D.13}$$

For one step of secondary driver:

$$\frac{d_3}{d_4} = \frac{\theta}{0.6°} \tag{D.14}$$

$$\therefore \theta = 0.3° \text{ if } d_4 = 2d_3 \tag{D.15}$$

Similarly

$$\tau'_{follower} = \frac{d_4}{d_3}\tau'_{driver} \tag{D.16}$$

$$= 2\tau'_{driver} \tag{D.17}$$

$$\text{But } \tau'_{driver} = \tau_{follower} \tag{D.18}$$

$$\therefore \tau'_{follower} = 2 \times 3\tau_{driver} \tag{D.19}$$

$$= 6\tau_{driver} \tag{D.20}$$

$$\therefore \tau'_{follower} = 6\tau_{driver} \tag{D.21}$$

Hence, we can conclude that using a system of pulleys, the angle is 6-fold smaller than the angular displacement of the motor and the corresponding torque is 6 times more than that specified for the stepping motor.

D.2 Torque Characteristics

Since in our system, one of the motors will orient the parabolic reflector, we will derive the necessary relationship between the torque of the motor and the centre of mass of the parabola. Considering a small ring of radius r, height dz and density σ as shown in Fig. D.3

$$dm = \sigma \, dA \tag{D.22}$$

$$= \sigma \, (2\pi r) \, dz \tag{D.23}$$

$$= \sigma \left(2\pi \left[\frac{z}{a} \right]^{\frac{1}{2}} \right) dz \tag{D.24}$$

since

$$z = ar^2$$

The centre of mass, $\mathbf{R}$ therefore is,

$$\mathbf{R} = \frac{1}{M} \int z \, dm = \frac{\int_0^z \sigma \times z \left(2\pi \left[\frac{z}{a} \right]^{\frac{1}{2}} \right) dz}{\int_0^z \sigma \left(2\pi \left[\frac{z}{a} \right]^{\frac{1}{2}} \right) dz} \tag{D.25}$$

$$= \frac{3}{5}z \tag{D.26}$$

Fig. D.3 Diagram showing the position of the ring being considered

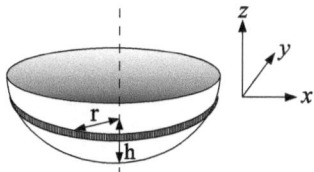

Fig. D.4 Torque required
from the axis of rotation of the
parabolic dish

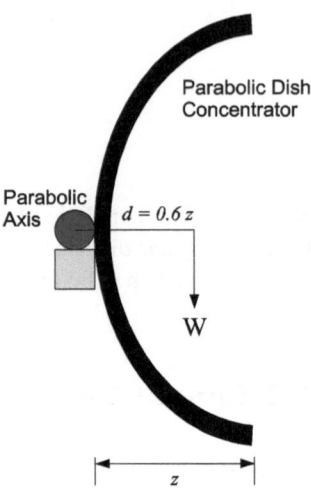

Now if we assume that the parabolic concentrating dish has a depth of 15 cm and
a mass of 5 kg, then the maximum possible torque as depicted in Fig. D.4 will be
7.5 Nm. However, if this dish is connected to a pulley system as in Sect. D.1, then a
motor of holding torque of only 1.25 Nm (173.6oz-in) is required.

Index